CHASING *the* STARS

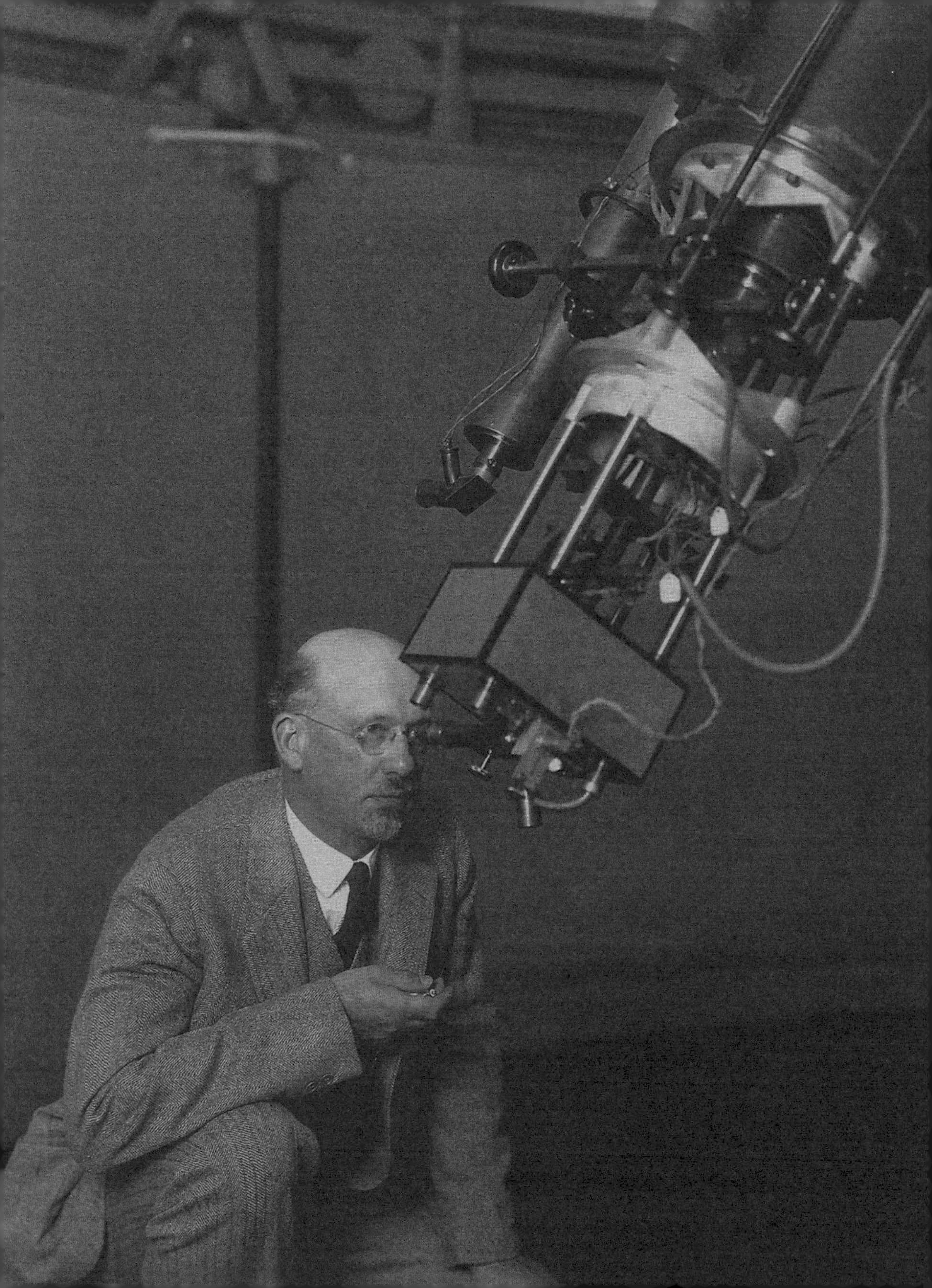

CHASING *the* STARS

How the Astronomers of Observatory Hill
Transformed Our Understanding of the Universe

JAMES LATTIS • KELLY TYRRELL

Published by the Wisconsin Historical Society Press
Publishers since 1855

The Wisconsin Historical Society helps people connect to the past by collecting, preserving, and sharing stories. Founded in 1846, the Society is one of the nation's finest historical institutions.
Join the Wisconsin Historical Society: wisconsinhistory.org/membership

Images on pages v, 4, 78, 156, 230, 235, and 238 are courtesy of Denny Moutray Photography, dennymoutrayphotography.smugmug.com.

Washburn Building on front cover: University of Wisconsin–Madison Archives, Madison, Wisconsin

Back cover images: Hubble Space Telescope (top), courtesy of NASA; astronomer Betty Webster (center) and Washburn director Joel Stebbins (bottom), courtesy of the Department of Astronomy, University of Wisconsin–Madison.

Printed in Canada
Cover design, text design, and typesetting by Steve Biel Design

28 27 26 25 24 1 2 3 4 5

Library of Congress Cataloging-in-Publication Data

Names: Lattis, James M., author. | Tyrrell, Kelly, author.
Title: Chasing the stars : how the astronomers of Observatory Hill transformed our understanding of the universe / James Lattis, Kelly Tyrrell.
Description: Madison, WI : Wisconsin Historical Society Press, [2024] | Includes bibliographical references and index.
Identifiers: LCCN 2023057030 (print) | LCCN 2023057031 (ebook) | ISBN 9781976600241 (paperback) | ISBN 9781976600258 (epub)
Subjects: LCSH: Washburn Observatory—History. | Astronomy—Wisconsin—History.
Classification: LCC QB82.U62 L38 2024 (print) | LCC QB82.U62 (ebook) | DDC 522/.1977583--dc23/eng/20240104
LC record available at https://lccn.loc.gov/2023057030
LC ebook record available at https://lccn.loc.gov/2023057031

♾ The paper used in this publication meets the minimum requirements of the American National Standard for Information Sciences—Permanence of Paper for Printed Library Materials, ANSI Z39.48-1992.

To all who have ever looked up to the sky with wonder.
And to the kind, ingenious, and dedicated people of Wisconsin's
Space Astronomy Laboratory. *Ad astra pro astris!*

CONTENTS

PART TWO

PART THREE

Washburn Observatory
PHOTO COURTESY OF JOHN RUMMEL

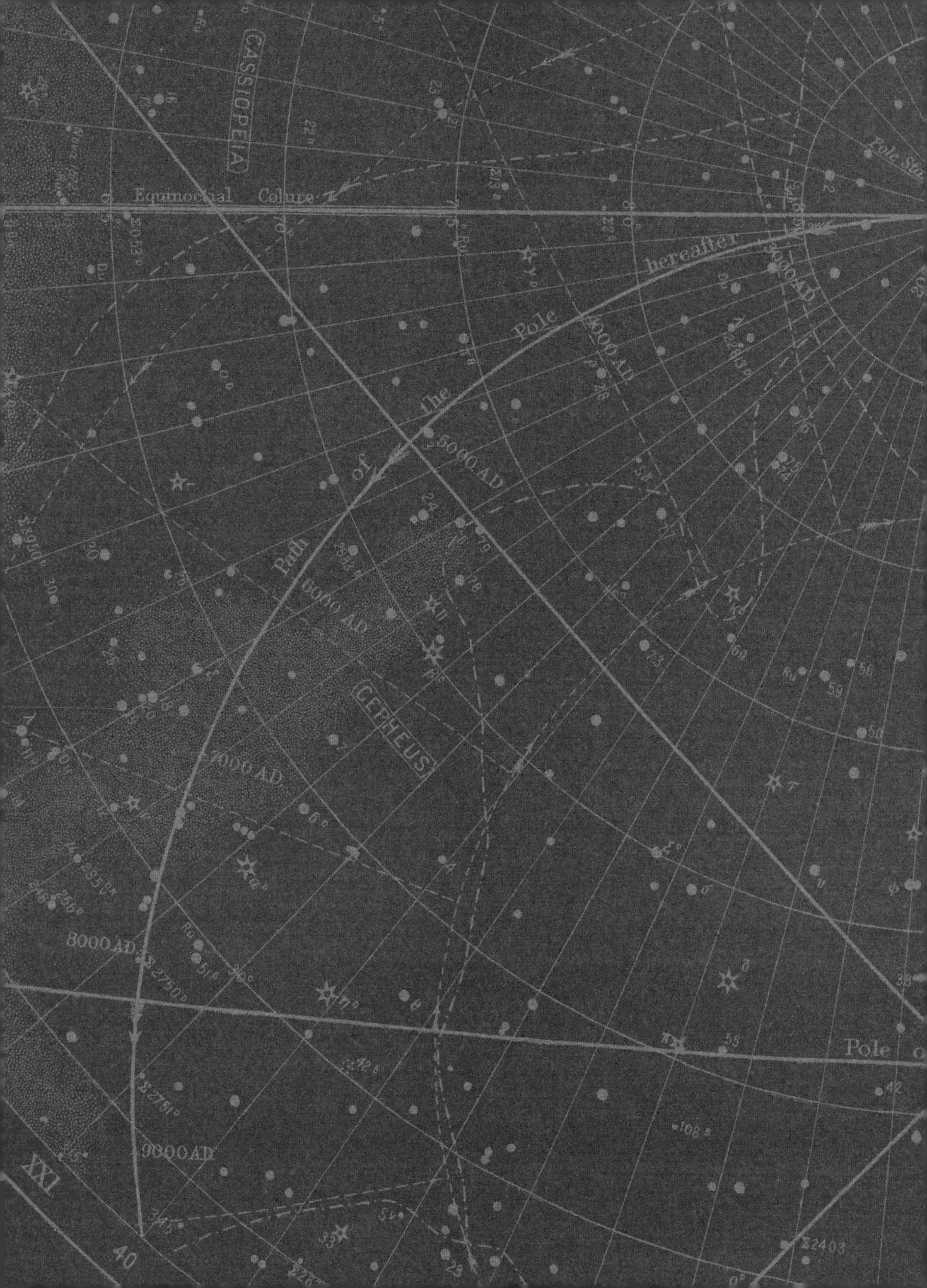
CASSIOPEIA
Equinoctial Colure
Path of the Pole hereafter
3000 AD
4000 AD
5000 AD
6000 AD
7000 AD
8000 AD
9000 AD
CEPHEUS
Pole Star
Pole o
XXI
40

CHASING *the* STARS

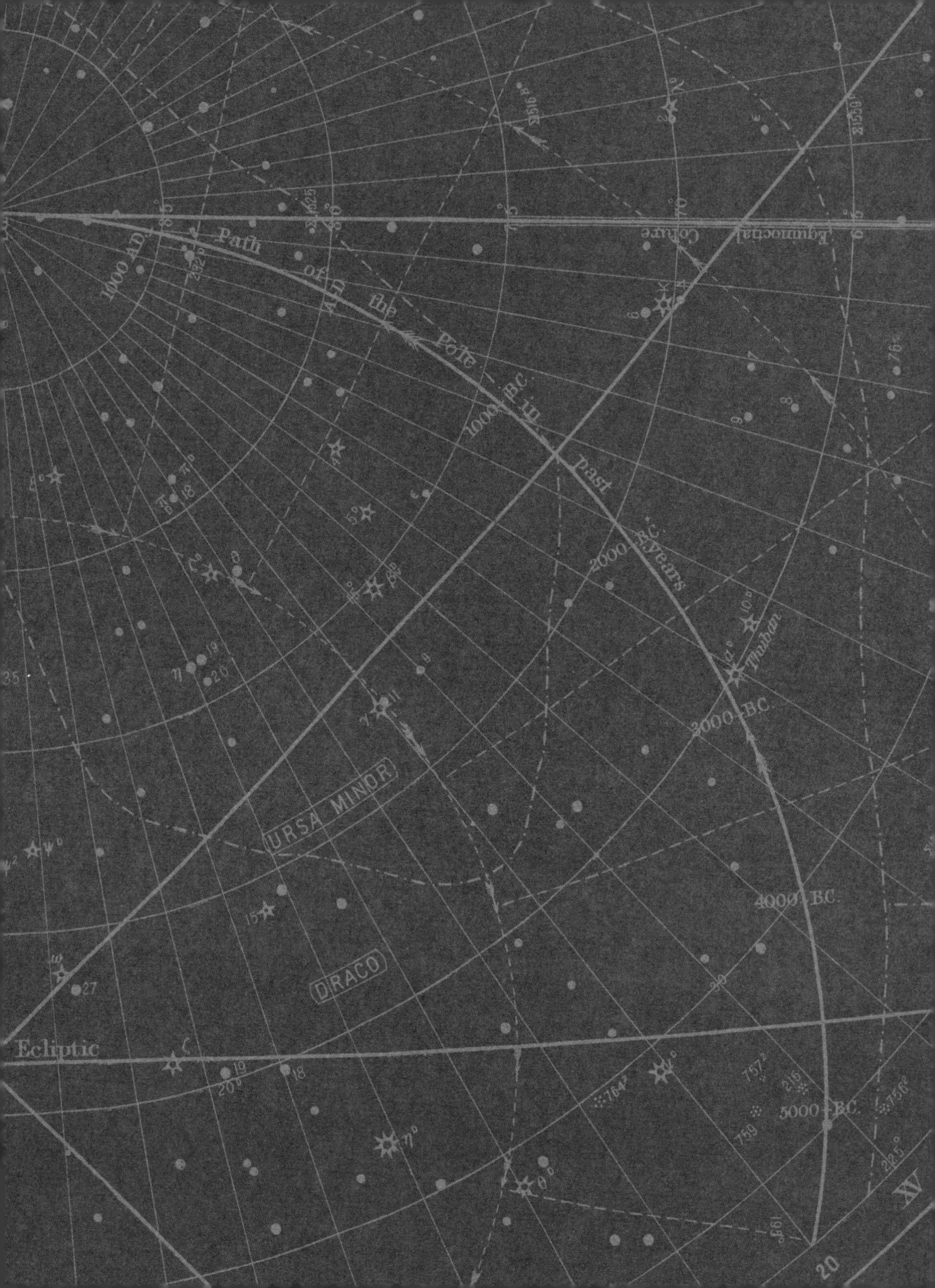

Path of the Pole in Past Years
1000 A.D.
A.D.
1000 B.C.
2000 B.C.
3000 B.C.
4000 B.C.
5000 B.C.
Thuban
URSA MINOR
DRACO
Ecliptic
Equinoctial Colure
XV
20

INTRODUCTION

This book explores the history of Washburn Observatory and the wellspring of innovation that developed there, ultimately leading to astronomical instruments flying in space and at work today in some of the world's largest telescopes.

Washburn astronomers were not the first stargazers in the land we now call Wisconsin, which has been inhabited for more than twelve thousand years. All human cultures regard and relate to the sky, day and night, in various ways. The earliest inhabitants of this land were no different, and the earthworks and artifacts of the Mound Builders and Woodland Indians reveal some of their astronomical knowledge.

Nor was Washburn the first observatory in Wisconsin when its foundations were laid in 1878 on the University of Wisconsin–Madison's Observatory Hill. Milwaukee College (later Milwaukee-Downer College) had a small observatory by about 1875, and others soon followed, including the long-vanished observatory of Sacred Heart College in Prairie du Chien, which was founded in 1882 by the Jesuit astronomer Johann Hagen, who later directed the Vatican Observatory. Beloit College had its own Smith Observatory by 1882; Lawrence University, in Appleton, established its Underwood Observatory in 1891; and the Carthage College Observatory, in Kenosha, was founded in 1893. Probably the most famous of all is the Yerkes Observatory, founded by the University of Chicago in 1897 in Williams Bay, Wisconsin, on the northern shore of Geneva Lake and less than ten miles from the Illinois border. Though located in Wisconsin, and while many of its astronomers were indeed Wisconsin residents, Yerkes Observatory long relied on Chicago's community and institutions for leadership. In 2018, Yerkes officially became a Wisconsin institution when the University of Chicago donated the buildings, instruments, and much of the surrounding property to the Yerkes Future Foundation, a Williams Bay nonprofit dedicated to preserving and developing the observatory as a historical and educational facility.

In contrast to the primarily instructional nature of other Wisconsin observatories (Yerkes aside), Washburn Observatory was the only one established as a research institution. While we do not neglect Washburn Observatory's role in campus and community life, the innovative research that made it famous in its time is what guides this historical account. This is not a scholarly history of Wisconsin astronomy in the formal sense, with the meticulous citation of primary sources or dialog with academic literature; instead, this book offers an accessible account of some of the most interesting and significant moments and developments from a distinguished yet little-known history. Our account is nevertheless based on original research from a wide range of source material.

Because we cannot fit the entire history of astronomical research at the University of Wisconsin, or even the full history of Washburn Observatory itself, into a single book, the arc of our story follows Wisconsin astronomy from its first days through the early twenty-first century, across which we can discern significant developments. We must work within the limitations of the historical record; as such, birth and death dates remain unknown for some of the people whose stories are included here, and initials are occasionally used when we could not confirm a person's first name. In some instances, all we have are oral traditions, passed down among people for generations.

The historical importance of innovative Wisconsin-based astronomical research from the most recent two decades remains to be studied. Worthy of note, however, is Wisconsin's partnership in the Southern African Large Telescope (SALT) project—a groundbreaking international effort to develop the largest single-optical telescope in the Southern Hemisphere, which officially opened in 2005. Wisconsin's active role in SALT's development was made possible thanks to former UW–Madison chancellor John D. Wiley, who in 2002 also created the Astronomy Department Board of Visitors, which has led the way in strategic planning and made possible major initiatives in astronomical research, education, and outreach.

The active research programs of astronomical significance happening in Wisconsin today are not limited to UW–Madison's Department of Astronomy, the direct descendant of Washburn Observatory. They span multiple UW–Madison departments, notable among them the Department of Physics, as well as other campuses of the Universities of Wisconsin, including those in Milwaukee and Whitewater. These programs and collaborations are far too numerous to be adequately summarized here, and they entail many fascinating and important histories that we must mostly pass over. Instead, we look back to the beginning, as we turn to the astronomical history that started on Observatory Hill.

PART ONE

"Whatever may be the limitations which trammel inquiry elsewhere, we believe that the great State University of Wisconsin should ever encourage that continual and fearless sifting and winnowing by which alone the truth can be found."

University of Wisconsin Board of Regents, 1894

1

EARLY STARGAZERS

THE MOUND BUILDERS

In the oral tradition of the Dakota, the Creator shaped the people from clay at the confluence of the Mississippi and Minnesota Rivers, after their spirits traveled to Earth along a path of stars in the Milky Way. The river waters converge and connect with earth and sky near the White Cliffs of what is now called Indian Mounds Regional Park, in St. Paul, Minnesota.

This sacred place holds special significance to the Dakota, alongside other Indigenous peoples of the Upper Midwest whose oral traditions tie them here, including the Ho-Chunk Nation of Wisconsin and the Iowa Tribe of Kansas and Nebraska. Over a period of more than one thousand years, beginning around 200 BCE, their ancestors built burial mounds atop these cliffs, which tower two hundred feet above the Mississippi. The Indigenous people who built the mounds lived in communities along the river, and from the high point on the bluffs, they mapped out the stars. Dakota scholars have identified correlations between the arrangement of the earthly mounds and celestial constellations. There were as many as two hundred mounds here before most were damaged, destroyed, or desecrated by European Americans in the nineteenth and twentieth centuries.

Farther north, on the face of a cliff that plunges into northern Minnesota's Hegman Lake, are the images, painted in red, of a human figure, a moose, and a small animal with a long, curved tail. For years, says Bois Forte Band of Chippewa (Ojibwe) member Carl Gawboy, people talked about these pictographs, and more than two hundred others along Minnesota's North Shore and Canada's Hudson Bay, "as mysterious, something from the past," and took guesses as to what they could be.

Gawboy, a retired professor of American Indian Studies at College of St. Scholastica in Duluth, spent years making countless sketches and photographs of the Hegman Lake figures, trying to "get into the heads" of the people who drew them hundreds of years ago. At the same time, he spent decades researching and asking questions about Ojibwe star knowledge. When he started to "think like a scientist" and not an artist, Gawboy realized the figures on the rocks represented Ojibwe constellations:

> *When I did that, that's when things started to go together. Who are the people that met there and said, "Well, this is what we have to remember, and this is what we have to teach, and this is how we're going to remember it, by putting these images on rocks: the Wintermaker, the Great Panther, and a Great Moose figure, so that we see the images in the rocks, we see the constellation and then there's the prophecy, the prediction, the story that goes with it—traditions that extend all throughout Ojibwe lore."*

Indigenous traditional knowledge provides tremendous insight, but it can be difficult to know for certain the relationship Paleo-Indians had with the sky. Much of what Western archaeologists and anthropologists today understand about prehistoric Indigenous astronomy in the Midwest is indirect, inferred by modern scholars' study of Ohio Hopewell sites established between 200 BCE and 200 CE, and prehistoric sites at Cahokia in southern Illinois near present-day St. Louis and at Trempealeau in western Wisconsin.

According to Sissel Schroeder, archaeology and anthropology professor at the University of Wisconsin–Madison, Western scientists believe early Indigenous peoples across the Midwest developed their own sophisticated uses of astronomy, most notably related to observations of solstice and lunar alignments. For instance, a site now known as Woodhenge sits just west of the famed Monks Mound of Cahokia—North America's largest pre-Columbian city, inhabited by the mound-building Mississippian people between 600 and 1400 CE. Woodhenge consisted of a series of circles made of large wooden posts. Some archaeologists believe Woodhenge functioned as a solar calendar, marking sunrise and sunset at equinox and solstice. (Others have argued that Woodhenge may have been built to solidify the social and political power of the city's elite by providing them a visibly direct connection to the Sun, rather than—or in addition to—serving as a calendar.) Fifteen miles west of Woodhenge, on a Cahokian site known today as the Emerald Acropolis, structures appear to be aligned with the most extreme positions of the Moon at standstill and at the halfway point of the cycle. Solar and lunar festivals would have drawn people to Cahokia from hundreds of miles away.

Around 700 BCE, the Woodland Indians began to inhabit the land we now call Wisconsin. In addition to building effigy and burial mounds, they domesticated plants, built homes and other structures, and made pottery. Their mounds often took the shape of animals or

The most complex known system of American Indian mounds in Wisconsin was constructed at Aztalan, in southeastern Wisconsin, by Mississippian people about one thousand years ago. Some of the mounds and landscape features at Aztalan suggest alignments with astronomical events, such as the winter solstice sunrise, though experts cannot determine whether the alignments were intentional. Here, a group of sixth-grade students visits Aztalan with UW anthropology professor Sissel Schroeder. JEFF MILLER, UNIVERSITY COMMUNICATIONS

Earthwork mounds constructed by Indigenous people are found widely throughout Wisconsin. This mound in Trempealeau, shown here in a photograph taken in 1905, overlooks the Mississippi River in the west-central part of the state. WHI IMAGE ID 114205

other figures. Around 1000 CE, Mississippians from Cahokia began to migrate to the region. At Aztalan, in present-day Lake Mills, Wisconsin, the Woodland Indians and Mississippians forged community within wooden palisade walls, merging pottery and housing styles and building additional mounds.

In the eleventh or twelfth century, the Cahokians appear to have reconstructed a sandstone ridge spur at Trempealeau, known today as Little Bluff. While scientists have not discerned a clear lunar or solar orientation to the reconstruction, they recognize the importance of other spiritual connections the people of Cahokia observed in relation to earth, water, and sky that could account for Little Bluff.

However, while astronomical alignments have been well-documented at some prehistoric sites and some Indigenous cultures have well-known connections to the celestial realm, it's important to resist making assumptions about why sites like Little Bluff were created.

When building their mounds across Wisconsin and other parts of the Midwest, the Woodland Indians favored elevated locations that often overlooked bodies of water. This means the summit we now call Observatory Hill was a prime location. Two mounds remain on the hill today. One of them, which has been identified as a two-tailed water spirit, is revealed here as the snow melts. The building in the immediate background is the northern end of Agriculture Hall. WHI IMAGE ID 5569

"Given the myriad of possible sighting points and lines among the thousands of differently oriented mounds and their appendages, almost any alignment that one is looking for could, eventually, be discerned," writes former Wisconsin state archaeologist Robert Birmingham in his 2009 book, *Spirits of the Earth: The Effigy Mound Landscape of Madison and the Four Lakes*.

In fact, ancient Indigenous people across the region had complex societies and built mounds and other structures within the context of varied social, political, and geographic factors. Though Indigenous people today carry the oral histories and traditional knowledge of their ancestors despite decades of forced removal from their lands by European American settlers, much about the Midwest's first stargazers remains unsettled.

What is certain, however, is that in building Washburn Observatory, at least one effigy mound on the land occupied by the university was destroyed.

Astronomer Edward Holden, who became Washburn's second director in 1881, bore witness to this history when he made the only known public acknowledgment that the observatory "was built upon the spot formerly occupied by an Indian mound." In the first volume of the *Publications of the Washburn Observatory*, published in 1882, Holden emphasized the importance of carefully preserving those mounds that survived.

"These are striking memorials of a race whose language even is unknown, and they serve to show the kind of antiquities which the young communities of the West afford," he wrote. "They should be sacredly preserved for present and future study."

2

OBSERVATORY FOUNDERS

WASHBURN, WOODMAN, AND BASCOM

In 1844, Cadwallader Colden Washburn (1818–1882) traveled from his native Maine to the Wisconsin Territory, where he soon went into business as a land agent at Mineral Point with fellow Mainer Cyrus Woodman (1814–1889). Together, Washburn and Woodman maintained a lucrative land and banking business for eleven years and helped establish civic infrastructure in the early days of Wisconsin statehood: Woodman helped found the Wisconsin Historical Society, and both men served as regents of the University of Wisconsin.

Washburn and Woodman eventually dissolved their business, after which Washburn turned his efforts to heavier industry, including logging and milling. Washburn also entered politics, winning election to Congress in 1854, and with his enterprises ever more aligned with the Mississippi River, he acquired a residence in La Crosse in 1859. When the Civil War broke out, Washburn received a commission as a colonel and formed the Wisconsin Second Cavalry. He served for the entire war and by the end had earned the rank of major general. After the war, Washburn returned to Wisconsin, where he resumed his lumber business and built a flour mill, the first of several, in neighboring Minnesota. He eventually partnered with Minneapolis businessman John Crosby and made a huge fortune in the milling business.

Cadwallader Colden Washburn. UNIVERSITY OF WISCONSIN–MADISON ARCHIVES, MADISON, WISCONSIN

The Washburn-Crosby milling company later became General Mills, still a familiar name. Washburn was reelected to Congress in 1867, then elected governor of Wisconsin in 1871. When his bid for reelection failed in 1873, Washburn retired from politics and decided to make a major gift to the University of Wisconsin.

In 1874, John Bascom (1827–1911) left his professorship at Williams College in Massachusetts and arrived in Madison to become the sixth president of the University of Wisconsin. A Yankee like Washburn and Woodman, Bascom shared their ambitions for improvement and social progress. Bascom was an experienced educator whose progressive vision included co-education, or the practice of educating women and men at the same institution. Inspired by German precedents, he soon began a push to emulate the new institutional model of a "research university," similar to what was already being done at Johns Hopkins, Cornell, Harvard, and Michigan, among others.

John Bascom. WHI IMAGE ID 33717

To advance Bascom's agenda, an astronomical observatory was the perfect project: Wisconsin lacked one; it would improve teaching and advance scientific research; it was a potent status symbol; and former governor Washburn was willing to build it. In 1876, Bascom and Washburn convinced the Wisconsin Legislature to provide state funding of three thousand dollars per year to support and maintain an observatory, "provided that such an Observatory shall be erected within three years on the University Grounds, without expense to the State." Washburn fulfilled this condition himself by building and equipping the observatory at the cost of $42,292.27 (as calculated later by the observatory's second director, Edward Holden). The new observatory was the first university building to be donated by a private individual and the first one dedicated to research.

Governor Washburn was closely involved in the observatory project: hiring the architect—David Jones of Madison—and the masons and carpenters; ordering the equatorial telescope from Alvan Clark and Sons; specifying and contracting for the major extension of the building that James Watson wanted; and, of course, paying for it all, including some salaries the university could not afford. Washburn even selected the site himself, pacing it off at the invitation of the Board of Regents.

As the observatory neared completion in 1881, Washburn's health began to fail (reports from the time sometimes referred to a kidney ailment known as Bright's Disease). Yet even through his decline, Washburn worked with Edward Holden to purchase the observatory's second major research telescope, a meridian circle instrument made by Repsold and Sons of Hamburg, Germany. Washburn had decided to visit European mineral water spas in search of treatment, so in August 1881 he added a stop at the Repsold shop in Hamburg to negotiate the terms of the meridian circle purchase in person. Unfortunately, the European waters provided only temporary relief. After Washburn's return to Wisconsin in early 1882, his health began to deteriorate. Washburn died on May 15, 1882, in Eureka Springs, Arkansas, where he had traveled for the spring waters. Washburn's meridian circle telescope, the final major component of his observatory project, would not arrive in Madison until that fall.

Cyrus Woodman. UNIVERSITY OF WISCONSIN–MADISON ARCHIVES, MADISON, WISCONSIN

Woodman, Washburn's former business partner, had moved to Cambridge, Massachusetts, in 1864, though the two men remained close friends. Woodman occasionally invested in Washburn's business ventures and, many years later, would serve as an executor of Washburn's estate. After Washburn's death, Woodman created from his personal wealth a trust fund with an initial investment of five thousand dollars to pay for books and journals, thus establishing the Woodman Astronomical Library, which for many years was housed in the Washburn Observatory itself. Under careful management of its trustees, the Woodman Trust still exists today and helps support the research and teaching needs of Wisconsin's astronomers.

3

WASHBURN'S FIRST TELESCOPE

THE CLARK 15.6-INCH REFRACTOR

When university president John Bascom announced in September 1877 that former governor Cadwallader Washburn was going to build an observatory for the University of Wisconsin, he proclaimed it would "be furnished with a fifteen-inch equatorial [telescope], 'equal or superior to that of the Observatory of Harvard University, Cambridge.'" We don't know who Bascom was quoting, but the invocation of Harvard might have been Governor Washburn's own idea of a criterion of excellence (possibly influenced by Washburn's friend Cyrus Woodman, a Harvard man). Since the telescope indeed would produce world-class scientific results for the next eighty years, it would seem to have been a good criterion.

"Equal or superior to" Harvard's was also a way of specifying a minimum size to Massachusetts telescope maker Alvan Clark and Sons. Harvard's equatorial refracting telescope, made by the German firm Merz and Mahler, had been installed in 1847, at which time its 15-inch lens diameter (the parameter always used by astronomers for the size of a telescope) was tied with the Russian Pulkovo Observatory as the largest in the world. However, by 1878, the world's largest telescope was the 26-inch refractor of the US Naval

OPPOSITE: The tube of the Clark 15.6-inch refractor, formed from riveted steel plates, has a focal length of 6,172 millimeters, so from lens to eyepiece it is more than twenty feet long. To see through the telescope's eyepiece, an astronomer at times must climb a large set of wooden steps that rise about thirteen feet above the floor.

This photograph from the mid-1920s shows the Washburn telescope still on its original Clark mounting. The iron mounting rests on top of a masonry pier that widens as it goes through the wooden floor, down through the rest of the building, and into the earth below. The rectangular box at the center of the mounting's base is the driving clock. Hanging on the end of the telescope is a first-generation Stebbins photoelectric photometer, shrouded by a dark jacket, and below it hangs a string electrometer. Next to the pier are the desk, chair, notebook, and lamp for the data recorder, who wrote down the electrometer indications read aloud by the telescope operator. WHI IMAGE ID 22332

Observatory in Washington, DC (also made by Clark). So, while Governor Washburn wasn't aiming for a record-setting size, at 15.6 inches, the first telescope of the new Madison observatory would nevertheless be the third largest in the country, after the 18.5-inch telescope at Chicago's Dearborn Observatory (also by Clark) and nosing ahead of Harvard's by half an inch.

George B. and Alvan G. Clark, sons of the firm's founder, personally delivered and installed Washburn's telescope in the new observatory dome in January 1879. Astronomer James Watson, who would soon become Washburn Observatory's first director, came to Madison to witness the telescope's installation and within a few days was testing the new instrument and hosting curious visitors. By late February both houses of the state legislature had visited the still-unfinished observatory to observe with the new telescope, and a few legislators returned on March 2 to observe the close approach that evening of Venus and Saturn.

During Watson's short tenure in Madison, he seems to have used the new telescope only for important visitors and for his astronomy class. But Washburn Observatory's second director, Edward Holden, with fellow astronomer Sherburne Burnham, used it for "sweeps" while searching for double stars (or stars revealed to be very close groupings of two or more stars) and new nebulae. Double star research with the 15.6-inch telescope continued for decades under the observatory's third director, George Comstock. By the 1920s, director Joel Stebbins and his assistant astronomer Charles Morse Huffer put the telescope to work with a succession of photoelectric photometers pursuing several research problems, the most enduring of which was the measurement of eclipsing variable stars—in a sense a continuation of the earlier double star work. In fact, Huffer continued eclipsing variable photometry almost until the telescope was retired from research in 1958.

In 1933, the aging telescope received a major renovation in the form of a new and improved mounting, which allowed it to continue operating for another twenty-five years. In total, Washburn's first telescope served as a research instrument for almost eighty years—one of the longer research careers of any classic refractor. In addition, Washburn's telescope has had an enduring influence on countless visitors and the many thousands of students from elementary school through university who, since 1881, have observed the Moon, a planet, a comet, or a double star on a clear night when it's dark on Observatory Hill.

In March 1933, work began on remounting the Clark 15.6-inch refractor. First the telescope tube would have been removed and stored. Then, pictured here, the components of the original Clark mounting, dating to 1879, were removed through the dome opening by crane. The components of the new mounting would later go in the same way, and the telescope was back in operation by June. UNIVERSITY OF WISCONSIN–MADISON ARCHIVES, MADISON, WISCONSIN

The new mounting for the Washburn telescope featured a much improved and more reliable clock drive, partially visible through the glass door on the pier, as well as several amenities to make the telescope operator's task easier. These included the right ascension and hour angle dial, visible on the north face of the pier, as well as a spoked wheel, like a ship's helm, just visible at the bottom of the image beneath the dial. This wheel allowed more convenient maneuvering of the telescope and made the various ropes of the old mounting unnecessary. DEPARTMENT OF ASTRONOMY, UNIVERSITY OF WISCONSIN-MADISON

4

WASHBURN'S FIRST ASTRONOMER

JAMES WATSON

James Craig Watson (1838–1880), born into a poor family in southern Ontario, rose to international fame in his short life. He was a bona fide boy genius whose talents became apparent after the family moved to Ann Arbor, Michigan, where his mother hoped that proximity to the University of Michigan would prove an advantage—a gamble that paid off. After studying the steam engine in the factory where his father worked, thirteen-year-old James became the company engineer when he proved his adult predecessor to be incompetent.

The Watson family was again impoverished when the factory closed, but his former employer interceded to keep James in school, where he tutored his schoolmaster in algebra and trigonometry. He entered the university at the age of fifteen, after mastering Greek and Latin on his own while becoming a skilled machinist. Watson soon came under the influence of German immigrant Franz Brünnow, director of the university's Detroit Observatory (and the first PhD astronomer in the United States). Watson immersed himself in theoretical and practical astronomy as Brünnow's only student and soon as observatory assistant. Then, in 1863, he became Brünnow's successor.

Preparation of star charts, combined with his prodigious photographic memory, led to the work that first brought Watson fame: the discovery of asteroids. By 1868, at the age of thirty, Watson had discovered nine of them. He soon was elected to the National Academy of Sciences and published his textbook, *Theoretical Astronomy*, which received both acclaim and notoriety for its enthusiastic borrowing from other authors. Watson would eventually discover twenty-two asteroids, a then-unprecedented number for one observer.

In parallel to his astronomical life, Watson worked as an actuarial consultant and then an investor in the life insurance business, which allowed him to accumulate a level of wealth usually unheard of for astronomers. His wife, Annette Waite, often assisted with his observing, and the couple traveled widely, taking a solar eclipse expedition to Sicily in 1870; visiting China (where he discovered another asteroid) for the transit of Venus in 1874; and, perhaps most famously, traveling to Wyoming for the solar eclipse of 1878, from which Watson returned convinced he had discovered the long-sought planet Vulcan, which was thought to orbit even closer to the Sun than Mercury. But his announcement immediately attracted serious criticism from other observers who had tried and failed to spot the mystery planet.

James Craig Watson. UNIVERSITY OF WISCONSIN–MADISON ARCHIVES, MADISON, WISCONSIN

It was at just this time that former governor Cadwallader Washburn was looking for an astronomer to direct the University of Wisconsin's new observatory in Madison. Governor Washburn wanted the best, and Watson was making headlines. President John Bascom and the Wisconsin Board of Regents engaged in a bidding war with the University of Michigan over the world-famous Watson, who finally accepted Wisconsin's offer, probably because of the new and superior equipment and the opportunity to build a new institution that could vindicate his controversial discovery of Vulcan.

Watson officially began his duties as observatory director in the summer of 1879 and immediately began shaping it to his purposes. From Ann Arbor he brought a student, George Cary Comstock, who would later serve as Washburn Observatory's third director. From Watson's own funds he began construction of two new buildings: a small Student Observatory, allegedly to deflect students away from the main instruments, and a small structure he called the Watson Solar Observatory, which he intended to use to prove that his Vulcan discovery was not the mistake it was beginning to look like—and, with additional observations, even determine the orbit of the new planet.

Watson also lobbied for major changes to the still-unfinished main observatory building itself, which was originally planned as a central dome with two small, symmetrical wings to the east and west. He convinced Governor Washburn that the main building should be expanded with an eastern extension that would contain a library, office space, sleeping rooms, and an instrument shop. Washburn promptly had his architect draw up plans, and construction began. Then, Watson sold Washburn on the idea of a major westward extension that could be used to house a large meridian circle telescope, also to be made by Alvan Clark and Sons, which would be too large to fit in the existing west wing (though the western addition was never built, owing to Watson's early death). A skilled instrument builder, Watson also began work on several new telescopes, including a large reflecting telescope that would have been a novelty in those days of large refractors. Amid his labors of organizing and improving the new observatory, not to mention teaching, it is no surprise that he accomplished no astronomical research in his short time in Madison.

Watson did not live to see any of his new projects completed. He taught astronomy in the academic year 1879–1880, during which he acquired a reputation as an affable teacher, interesting lecturer, and easy grader. For example, one night in June, he opened the observatory to the students of the senior class and observed with them until four o'clock in the morning, as reported in the *Wisconsin State Journal*. Watson and Annette had resided as guests of Governor Washburn at his Edgewood estate upon their arrival in Madison but moved into the Observatory Residence in late summer 1880. Watson enjoyed his hilltop home for only a few weeks. As he worked energetically on his projects in fall of 1880, his health suddenly declined and he collapsed, perhaps while working on the basement of his Solar Observatory. Watson died on November 22, 1880, owing to either peritonitis, pneumonia, or a "congestive chill," depending on the source. Watson's death, which news reports described as a "calamity," was widely felt across Wisconsin and the nation. A delegation of University of Wisconsin faculty members accompanied his body on the train to Ann Arbor, where he was buried.

While his claim of discovering Vulcan would later be disproved, this does not diminish Watson's importance as a member of the first generation of American astronomers to establish the nation's reputation for scientific accomplishment. Watson was the first in a long line of innovative instrument builders among Wisconsin's astronomers, and his Student Observatory eventually would serve as an important research tool. In his short tenure, Watson established a strong scientific tradition that helped bring about important developments in astrophysics and space astronomy and that flourishes to this day.

5

THE VERSATILE STUDENT OBSERVATORY

As director of the University of Michigan's Detroit Observatory in Ann Arbor, James Watson had a reputation for being stingy with telescope time, keeping it to himself rather than allowing access to university students and the public. When Watson arrived in Madison as Washburn Observatory's first director, he promptly began building, at his own expense, a Student Observatory, where astronomy students could learn how to operate astronomical instruments and gain experience observing the sky without getting in the way of his research with the 15.6-inch Clark refractor. Like all of Watson's projects, the Student Observatory was still unfinished when he suddenly died in 1880, but Governor Washburn finished the new building with the help of student engineer Frank Winkley, who designed the mechanical details of the dome and roof openings.

Watson's plan for the Student Observatory called for a small equatorial refractor and dome connected to a west wing that would house a transit instrument. This was the classic astronomical duo in that era for a well-equipped observatory. The transit house included a slit opening to the sky running north to south in the roof and walls so that the transit instrument, a telescope mounted to pivot on a single horizontal axis, could observe anywhere along the local meridian. Less typically, and presumably prescribed by Watson, Winkley's design of the transit house included a slit opening along the peak of the roofline (which ran east to west), with matching shutters in the east and west walls, thus allowing observations along the prime vertical (the celestial great circle crossing through the zenith from east to west). The dome, therefore, had to be slightly offset to the south of the east-west roofline of the transit house to avoid blocking the view of the eastern prime vertical. An anteroom, containing the entryway that connects the dome and meridian room, fits into that offset.

The first telescope in the dome was Sherburne Burnham's famous six-inch refractor, which he brought to Madison in the summer of 1881, when he arrived as Washburn's first

The Student Observatory, pictured here in the 1880s, stood at its original location until its relocation in 1960. In the background is Lake Mendota and, in the distance, Picnic Point. The dome for the equatorial telescope is on the east (right) side of the building, and the transit house is on the west (left). The roof of the transit house could open along the local meridian running north-south, or it could open along the east-west roof line for observations along the prime vertical. For this reason, James Watson's design offset the center line of the transit wing slightly northward of the centerline of the dome, so that the dome did not block the transit wing's view of the eastern sky. The dome was constructed from a wooden frame covered in metal sheeting. DEPARTMENT OF ASTRONOMY, UNIVERSITY OF WISCONSIN–MADISON

assistant astronomer. The Burnham telescope served off and on, for both research and teaching in the Student Observatory, until about 1931, when it gave way in the dome to a new ten-inch refractor. The transit house originally held a three-inch transit telescope made by Fauth and Co. in Washington, DC, and purchased by Governor Washburn. In 1897, Director George Comstock replaced it with what he called "the finest instrument of its type I have ever seen": a 76-millimeter "broken" transit (referring to the 90-degree reflection in the optical path) made by Carl Bamberg from Berlin, Germany, which was still in service there when director Joel Stebbins took over in 1922.

The Student Observatory was a significant research facility. Its dome underwent major modifications, as did the Burnham refractor, so it could play its most important role, in Comstock's 1889–1892 research project, measuring the aberration of starlight and atmospheric refraction, two essential phenomena that astronomers must allow for in careful

Sherburne Burnham's six-inch refractor was mounted in the Student Observatory by the fall of 1881, on loan at first, but purchased by the Board of Regents for the observatory in early 1882 for $1,200. Here it's seen on its original Clark mounting, with its base slightly shimmed up on the north (left) side to account for the latitude difference between Madison and Chicago, its original location. WHI IMAGE ID 22331

The 1893 Bamberg "broken" transit was purchased by Comstock for the transit house of the Student Observatory. The "broken" optical path (reflected through 90 degrees) made the instrument more compact and easier to use than a conventional transit telescope. UNIVERSITY OF WISCONSIN–MADISON ARCHIVES, MADISON, WISCONSIN

position measurements. Much later, in the 1930s, when the dome then housed the ten-inch refractor, it was used for testing and observations in photoelectric photometry.

In 1958, when the astronomers moved to Sterling Hall, equipped on its roof with observatories, and observers began working at the new Pine Bluff Observatory in nearby Cross Plains, the Student Observatory became entirely irrelevant. The Madison Astronomical Society, a club of mostly amateur astronomers, organized an effort to move the Student Observatory building from Observatory Hill to another location. A transfer of state property to a private group required the approval of the Wisconsin Legislature—which was granted through Act 297 of 1959. Funding for the building's relocation was donated by Oscar G. Mayer Jr., of the Madison-based meatpacking company. Mayer who happened to be an avid amateur astronomer, and the Madison astronomy club named their repurposed observatory after him in recognition. In July 1960, the club moved the building from Observatory Hill to a hilltop location a few miles away in Fitchburg, Wisconsin, on the grounds of Bjorksten Research Laboratory, where it was reoriented ninety degrees, with the originally western transit wing now north of the dome. The little building—under the name Oscar Mayer Observatory—became a Fitchburg landmark and symbol, even occasionally depicted on town police vehicles.

After ownership of the Student Observatory was legally transferred to the Madison Astronomical Society, the building was partially deconstructed in the summer of 1960 and transported from Madison by truck to its new location in Fitchburg. UNIVERSITY OF WISCONSIN–MADISON ARCHIVES, MADISON, WISCONSIN

Local amateur astronomers used the observatory until the mid-1980s, when increasing light pollution and the aging structure caused the Madison Astronomical Society to vacate the building for a more rural observing site they had acquired. The observatory was given new life yet again more than thirty years later, when Promega Corporation acquired the Bjorksten properties, including the observatory building. Beginning in 2018, Promega refurbished, remodeled, and reequipped the little observatory with a modern telescope, advanced controls, and excellent imaging capabilities. In 2022, Promega named the observatory in honor of astronomer Jocelyn Bell-Burnell and dedicated it to research, education, and public uses. Thus, Watson's original Student Observatory continues its astronomical and educational missions to this day, as it approaches one hundred fifty years of service.

6

CHASING VULCAN

THE WATSON SOLAR OBSERVATORY

The search for the planet Vulcan, once thought to orbit even closer to the Sun than Mercury, is a delightful example of how and why modern science works. Astronomers first hypothesized the existence of Vulcan as a means to account for the observed orbital motion of Mercury around the Sun. By 1859, they knew that the gravitational influences of known planets could not entirely explain Mercury's orbital motion. The gravitational tugs of the other already known planets came close to giving the right answer, so theoretical astronomers suggested that the discrepancy could be caused by an undiscovered planet orbiting close to the Sun. Several astronomers actually reported they had seen an unexpected planetary body crossing the disk of the Sun, and this suspected (though still unconfirmed) intra-Mercurian planet was given the name Vulcan, after the Roman god of fire, metalworking, volcanos, and the like, who is always near his hot forge.

Like a number of other astronomers of his day, James Watson hoped to confirm the existence of Vulcan and determine its orbit during a total solar eclipse by searching the darkened sky near the Sun during totality. During the eclipse of July 29, 1878, a year before his move to Madison, Watson believed he had discovered the mystery planet, and he announced the finding to the world. But of the many other astronomers searching for Vulcan during that eclipse, only one, Lewis Swift, also reported seeing an unknown object. Worse, Swift's report was inconsistent with Watson's. Watson's reputation as a record-breaking asteroid hunter lent considerable credibility to his claim, but many astronomers had grave doubts and were very critical. The finding would need to be independently confirmed.

As the new director of Washburn Observatory in 1879, Watson set out to vindicate his discovery, and, with more observations, even determine the orbit of the new planet. And he wouldn't need to wait for another total solar eclipse. His plan was to create a special instrument based on what was considered common knowledge at the time: that if viewed

This photograph of the Washburn Observatory complex from the 1890s shows the Watson Solar Observatory in the foreground. Visible on the left is a staircase connecting the Solar Observatory to the crest of the hill. A box-like structure at the top of the staircase houses the heliostat mirror. This structure is connected to the basement of the Solar Observatory by way of the polar axis tunnel, running below ground parallel to the staircase. At the far right, the outline of the Observatory Residence is just visible through the trees. WHI IMAGE ID 67856

through a long tube, such as a chimney or from the bottom of a well, it was possible to see stars and planets in the daylight sky. In fact, telescopes can show some objects, such as bright stars and planets, in the daytime, so the plan seemed to Watson worth a test. With his own funds he designed and began to build a small structure, which he called the Watson Solar Observatory, down the southern slope of Observatory Hill and slightly west of the observatory itself. The building had a deep basement, and from that basement a tunnel sloped northward and upward to emerge at the crest of the hill just west of the main building. The tunnel pointed up at the north celestial pole.

Watson's plan called for a specially mounted mirror, called a heliostat, at the top of the tunnel, which would rotate with the Sun's daily motion and reflect sunlight, or light from a patch of sky near the Sun, down the tunnel to a telescope in the solar observatory's basement. Looking through the telescope aimed skyward through his tunnel, Watson expected he would be able to see Vulcan near the Sun in the daytime sky and thus confirm its existence and motions.

But when Watson died suddenly in November 1880, the solar observatory was still unfinished. Watson's successor, Edward Holden, also believed the idea worth testing and successfully lobbied Governor Washburn for the funds to complete it. By the summer of 1882, Holden had finished construction (complete with housing at ground level for an assistant), borrowed a heliostat from his friend and fellow astronomer Samuel Langley, installed Burnham's six-inch telescope in the basement aimed up through the tunnel, and begun tests of Watson's concept.

The project was a total failure, revealing no stars in the daytime sky of brightness comparable to the putative Vulcan. Holden reported the disappointing results to the university and published them in a brief note in a German astronomical journal. The dank basement of the Solar Observatory proved to be a terrible place for a telescope, so the little building couldn't even be used for solar observations, which had been its secondary justification. The heliostat was returned to Langley, the six-inch telescope remounted in the Student Observatory, and the tunnel sealed up.

The Solar Observatory building did, however, continue to serve as lodging (at ground level) for an observatory assistant and later as a storage building. By the 1930s, the building had become an attractive hazard. In 1949, a demolition crew preparing to tear it down accidentally set it on fire, which brought about its rapid, uncontrolled destruction. Nancy Nicholas Hall now extends over the site. Construction of that building briefly revealed Watson's polar-aligned tunnel, which still points up toward the North Star from beneath Observatory Hill.

The only piece of the Watson Solar Observatory that survives today is the stone name plaque, visible here along the outer wall of the chimney directly below the roof line. DEPARTMENT OF ASTRONOMY, UNIVERSITY OF WISCONSIN–MADISON

Though futile, Watson's Solar Observatory was an ingenious attempt to test Newtonian gravitation by finding the mystery planet that theory seemed to predict. Other astronomers, including Holden, would continue searching for Vulcan during total solar eclipses as late as 1914, but to no avail: it doesn't exist. Albert Einstein showed in 1915 that his General Theory of Relativity could explain—as Newtonian gravitational theory could not—the seemingly anomalous motion of Mercury as a result of the strong curvature in spacetime near the large mass of the Sun. Despite its failures, the Solar Observatory was the first in a long line of innovative (and generally much more successful) astronomical instruments that Wisconsin astronomers continue to develop and operate to the present day.

7

WASHBURN'S SECOND DIRECTOR

EDWARD HOLDEN

A biographer who knew him well once called Edward Singleton Holden (1846–1914) "an extremely interesting man" whose "conversation was entertaining to the point of brilliancy." Holden became Washburn Observatory's second director in February 1881, taking up the post after James Watson's unexpected death. Governor Washburn had first met Holden—and had, perhaps, been charmed by his loquaciousness—in 1880, when Holden was a junior astronomer at the US Naval Observatory. Holden had already been consulting for some years in the development of California's Lick Observatory, which he hoped someday to lead, so he had many ideas about how to build and operate an astronomical observatory. To Governor Washburn, Holden looked like the perfect candidate to be the new director, and the university's regents accepted Washburn's judgment.

Edward Singleton Holden. UNIVERSITY OF WISCONSIN-MADISON ARCHIVES, MADISON, WISCONSIN

Holden, his wife, Mary Chauvenet, and their children moved into the Observatory Residence in the spring of 1881, and he quickly began bringing to fruition the efforts Watson had begun. He assembled and mounted the advanced sidereal clock that Watson had ordered from Anton Hohwü of Amsterdam, which would serve as the observatory's astronomical

Edward Holden wears his heavy fur winter-observing cloak while posing with Washburn Observatory's Repsold meridian circle telescope. Behind Holden hangs the lamp used to illuminate the scales of the circle when configured on the western end of the telescope. In the glass-sided box on the far left is a collimator telescope used to adjust the meridian circle, and next to it, on a four-legged stand, is the jack used to lift the meridian circle from its pivots. This allowed it to be rotated, or "reversed," 180 degrees, moving the circle between east and west configurations. In the foreground left of center is the reclining chair for the observer whose job was to note the moment of transit as seen in the eyepiece. In the foreground right of center is the bench for the second observer who would read the circle when configured (as here) in its eastern position (Holden is posing on the corresponding western bench). DEPARTMENT OF ASTRONOMY, UNIVERSITY OF WISCONSIN–MADISON

clock for decades. And he convinced Governor Washburn to fund the completion of the Student and Solar Observatories and then oversaw the work.

But Holden didn't simply follow Watson's lead. He advised Washburn to abandon Watson's plan for a large Clark meridian circle, which would have required a new western extension to the building, and instead convinced the governor to purchase a smaller meridian circle instrument from Repsold and Sons of Hamburg. That instrument would fit in the existing west wing and was less expensive, freeing up salary for an additional astronomer. The astronomer Holden had in mind was Chicago's famous double star observer, Sherburne Wesley Burnham. Although Holden was not a particularly gifted researcher himself, he knew the importance of recruiting top talent to build a scientific institution, and Burnham was only the first of those whom Holden would bring to Washburn (he would later follow the same pattern as director of Lick Observatory in California). Holden also convinced Burnham to sell to the university an instrument that would become one of Washburn's most important research tools: Burnham's own personal telescope, the six-inch Clark refractor, which Alvan Clark and Sons had made for him in 1870 (making it the university's oldest telescope, though not its first).

Holden made the observatory library a priority, designing and organizing it himself. He produced the observatory's first scientific publication, the first and only issue of the *Contributions of Washburn Observatory*, in May 1881, and began the *Publications of the Washburn Observatory*, a series that distributed the research of Washburn's astronomers until 1934, by which time modern astronomical journals were becoming the primary medium for scientific publication. Holden also established the observatory's time service, by which accurate astronomically regulated time signals were sent over telegraph lines to paying customers—most importantly railroads—bringing in a substantial portion of the observatory budget for some years.

One of Holden's most enduring accomplishments was to establish in April 1881 a regular schedule of public visitor nights at the observatory, held on the first and third Wednesday evenings of each month, founding a tradition honored to this day that is interrupted only by weather, equipment malfunctions, or public emergencies. And in 1883, at the invitation of the National Academy of Sciences, Holden led the US expedition to Caroline Island for the solar eclipse of May 6, the first of many eclipse expeditions led by the directors of Washburn Observatory.

In late 1885, after an eventful and successful tenure in Wisconsin, Holden left Madison to accept the office of president of the University of California, which was destined to receive the Lick Observatory once completed, with Holden as its designated director. In the summer of 1888, Holden finally moved to the top of Mount Hamilton as director of Lick Observatory, embarking on a complex and stormy career that eventually would stand in stark contrast to his productive years in Madison.

8

WASHBURN'S FIRST RESEARCHER

SHERBURNE BURNHAM

The first significant research done at Washburn Observatory was accomplished by an amateur astronomer who was far more than a hobbyist with a telescope. He was a stenographer and court reporter from Chicago named Sherburne Wesley Burnham (1838–1921). By 1881, Burnham was already internationally famous as a prolific discoverer of double stars, having outdone professional astronomers with his six-inch backyard telescope while also keeping his day job. He had met Edward Holden in 1874 at the US Naval Observatory, where Holden was an assistant astronomer and astronomical librarian, when Burnham visited to consult their double star catalogs.

When Holden became director of Washburn Observatory, he invited Burnham to come to Madison and pursue his double star work full time using the big, new Clark refractor, equipped with a fine Clark filar micrometer too expensive for Burnham to afford on his own. Because the university budget covered the salary for only one astronomer, Holden convinced Governor Washburn to personally pay Burnham's salary, two thousand dollars per year, as assistant astronomer. Burnham took temporary leave from his court reporting job in April 1881 and came to Madison to become a professional astronomer, only one month after Holden himself had arrived, thus doubling Washburn's astronomical staff, though unfortunately not for long.

Holden recognized the importance of demonstrating the scientific potential of the new observatory, both to fulfill Governor Washburn's generosity and to attract support from the university and state. So, in May 1881, their first month of observing, Holden and Burnham busily discovered twenty-nine new nebulae and sixty-two new double stars, work that Holden promptly documented in *Contributions of Washburn Observatory*, the pamphlet they rushed out that summer as the new observatory's very first publication.

Sherburne Burnham and Mary Cleland married in 1868 and had three sons and three daughters. They lived in Chicago, where Sherburne had moved just after the Civil War, for the rest of their lives, outside of occasional sojourns at observatories. COURTESY SPECIAL COLLECTIONS, UNIVERSITY LIBRARY, UNIVERSITY OF CALIFORNIA SANTA CRUZ. LICK OBSERVATORY PHOTOGRAPHS

Burnham obtained micrometer measurements of those newly discovered double stars, as well as another eighty-eight that he discovered during the rest of that summer. Burnham also applied his mechanical ingenuity to the Clark filar micrometer to improve the illumination of the wires seen by the observer in the eyepiece. He published the design of his modifications, which became widely adopted. The Clarks adapted Burnham's design for the micrometer they made for the Lick Observatory's 12-inch refractor, which Burnham would later put to use there.

By June, Burnham had pretty much decided that despite Holden's efforts to convince him to stay, he would not be taking a long-term position on the Washburn staff as long as his salary was funded only by Governor Washburn. The former governor's health was failing, and Burnham was worried about supporting his family in Chicago. Without an appointment backed by the university, he had little security, so he would eventually have to return to his job in Chicago as a court reporter for the US Circuit Court or risk losing it. Burnham kept up the observing in Madison through September 1881, after which he accompanied Holden to California to set up Lick Observatory's first telescope, a 12-inch Clark refractor, and with it observe the transit of Mercury. Holden returned to Madison, and Burnham went home to Chicago, leaving his famous six-inch Clark telescope behind in Madison.

Although Holden never convinced Burnham to work again at Washburn, after becoming director at Lick, Holden did eventually lure Burnham to join the staff there in 1888. That arrangement eventually ended badly by 1892, with Burnham acrimoniously resigning and once again returning to Chicago and his job as a court reporter. Burnham would later come back to Wisconsin to join the staff of Yerkes Observatory, much closer to Chicago than Madison.

Burnham made several modifications to the original Clark filar micrometer. The eyepiece is centered in the rectangular outline of the micrometer. The knob directly below the eyepiece adjusts the position of the movable wire in the micrometer, while the knob to the lower right of the eyepiece rotates the micrometer around the optical axis of the telescope to measure position angle.

Burnham, and later Comstock, used this micrometer extensively for double star measurements. Shown here, it is mounted on the eyepiece end of the telescope in its working position, but with a paper background hiding the telescope to make the micrometer structure easier to see. This photograph was used to produce engravings for publications in which Burnham explained his improvements, which were immediately adopted for the micrometer being built for Lick Observatory. So here we see the relatively crude prototype of Burnham's improved micrometer. To use such a micrometer for stellar measurements, the "wires" were illuminated from the side by an oil lamp. In conventional micrometers, the position of the lamp had to be adjusted nearly every time the micrometer was rotated or the telescope moved. Burnham created a way to mount the lamp, in the upper right, on a kind of gimbal so it remained vertical at all times. The light from the lamp was reflected by a mirror onto the wires so that as the lamp, micrometer, and telescope moved, the light always illuminated the wires at the correct angle.
DEPARTMENT OF ASTRONOMY, UNIVERSITY OF WISCONSIN–MADISON

SPOTLIGHT

THE OBSERVATORY'S FIRST PHOTOGRAPH, 1881

The earliest known photograph of Washburn Observatory was taken in the summer of 1881. It captures a westward view from the roof of the observatory director's residence. Although the photographer is unknown, the photograph must have been the project of Edward Holden, the second director of the observatory, who arrived in Madison in March 1881. One month later, the final section of the main observatory—the east extension, visible here facing the camera—was completed. The shadows indicate the photo was taken about a month before or after the summer solstice, in June or July.

Above and behind the east wing rises the dome of the 15.6-inch equatorial telescope, the covered lens of which points out of the dome slit. The identity of the man standing in the open slit, like the identities of others in the photograph, remains unknown. The dome shutter reveals that the photo must date to the summer of 1881, because by late 1881, Holden had replaced the lower shutter (visible here above the telescope in its open position) with a new design: a pair of doors that open outward, disconnected from the upper shutter. The dome opens this way today.

In the lower right is the Student Observatory. By the time this photo was taken, the six-inch telescope belonging to famed astronomer Sherburne Burnham was mounted in the dome. Burnham, who served on the Washburn Observatory staff during the summer of 1881, is probably one of the men in the photo. Barely visible through the tree branches and directly to the left of the main building at the edge of the photo is the bright roof of the Watson Solar Observatory, which was located slightly down the south slope of Observatory Hill.

In addition to the mystery man in the dome, one man can be seen sitting in a chair in the entryway of the main building's east wing. This could be Holden, though it is impossible to tell. At least two children—one on the steps just to the man's left and the other standing

The first photograph of Washburn Observatory dates from summer 1881, just after the completion of the observatory's main buildings. UNIVERSITY OF WISCONSIN–MADISON ARCHIVES, MADISON, WISCONSIN

at the rightmost corner of the main building—are evident in the photograph and appear to be about the ages Holden's children would have been at the time. The family lived next door in the Observatory Residence, on the roof of which the photographer was perched.

Among the unidentified men could also be Frank Winkley, the student engineer who designed and built the new dome doors; C. M. Conradson, a student assistant; or George Cary Comstock, a student assistant who was deeply involved in completing the observatory and who several years later would succeed Holden as director.

This photo inspired an iconic engraving that Holden commissioned for use in the *Publications of the Washburn Observatory*, and it indeed appeared in volumes one through four. However, Holden directed the engraver to excise the people and make some other changes: the steps up to the door of the Student Observatory are wider and slightly reoriented, and the six-inch telescope in the Student Observatory dome, which is nearly lost in the shadow in the photo, appears perfectly clear.

This engraving made from the first photograph appeared as the frontispiece in several early volumes of the observatory publications. DEPARTMENT OF ASTRONOMY, UNIVERSITY OF WISCONSIN–MADISON

9

THE FORGOTTEN TELESCOPE

THE REPSOLD 4.8-INCH MERIDIAN CIRCLE

One of the most complex research instruments in a nineteenth-century observatory was the meridian circle. It was essentially a telescope mounted on an east-west aligned horizontal axis, so it could point only along the local meridian, which is the arc across the sky from north to south passing through the zenith. The specialty of a meridian circle instrument was that it allowed astronomers to precisely determine the position of objects in the sky.

Reclining on a special chair beneath the eyepiece, an observer would watch a star traverse the telescope's visual field, crossing the meridian. The exact time of meridian passage could be used to measure the east-west position of the star (or other target object). Working with the observer, a second astronomer recorded the elevation of the transiting star by examining a very finely divided circular scale (the "circle") mounted on the horizontal axis. This provided a north-south position. These raw measurements were used later, generally in the daytime by the observatory's human computers, or by the observers themselves on a cloudy night, to convert the meridian circle measurements into the position of the target object among the stars.

The finely divided and inscribed circular scale on the Washburn meridian circle could be examined by four microscopes, one in each quadrant of the circle, for greater accuracy. Errors could be further reduced by lifting the entire telescope, axis, and circles away from the mounting piers on a special jack and rotating or "reversing" around the vertical axis, then repeating a measurement, which would cancel out some instrumental errors. Thus, the meridian circle was a very complicated instrument to make, adjust, maintain, and use.

Because it could point only along the local meridian, a meridian circle instrument did not need a dome. Instead, the west wing of Washburn Observatory was built with a series of shutters (some still visible today) that opened the walls and roof to a slice of sky stretching from north to south. The Washburn Observatory meridian circle was made by Repsold

Assistant John Tatlock stands in front of Washburn Observatory's Repsold meridian circle telescope. Tatlock was hired in 1883 as the first dedicated observer for meridian circle work. He would remain involved with the astronomical community but went on to a lucrative career in the insurance business.
UNIVERSITY OF WISCONSIN-MADISON ARCHIVES, MADISON, WISCONSIN

and Sons of Hamburg, Germany, and was similar to the Repsold meridian circles made later for Lick Observatory and Goodsell Observatory at Carleton College, in Northfield, Minnesota (where their instrument is preserved in nearly original condition today). Governor Washburn himself made the trip to Hamburg in the summer of 1881 to negotiate the final details—for example, that its 4.8-inch objective lens would be made by the American firm Alvan Clark and Sons—but he did not live to see it become the second major research instrument of his observatory. It took more than a year for the Repsold shop to complete and ship the meridian circle, which was mounted in the west wing in early December 1882; and student John Tatlock was hired in the summer of 1883 as observing assistant for the new instrument. According to Edward Holden, the total cost of the meridian circle, including shipping, was a little more than four thousand dollars.

Perhaps the Washburn meridian circle's most important project, involving many years of work that was mostly performed by astronomer Albert S. Flint, was the measurement of the slight shifts in star positions called stellar parallax, from which can be derived the actual distances between stars and the Earth. The Washburn catalog was an important source for parallaxes of stars in the early twentieth century. According to Joel Stebbins, back when parallaxes were known for only about one hundred stars, half of those had been determined by Flint at Washburn Observatory.

By the 1920s, photographic plates began replacing instruments like the meridian circle for the measurements of star positions. A photographic plate recorded relative positions of many stars in a single observation, the exact positions of which would be worked out later. Once perfected, the photographic method was far more efficient than the one-star-at-a-time measurements of the meridian circle. As a result, the complex meridian telescopes rapidly became obsolete. When Flint retired in 1920, the Washburn meridian circle retired with him and was never used again.

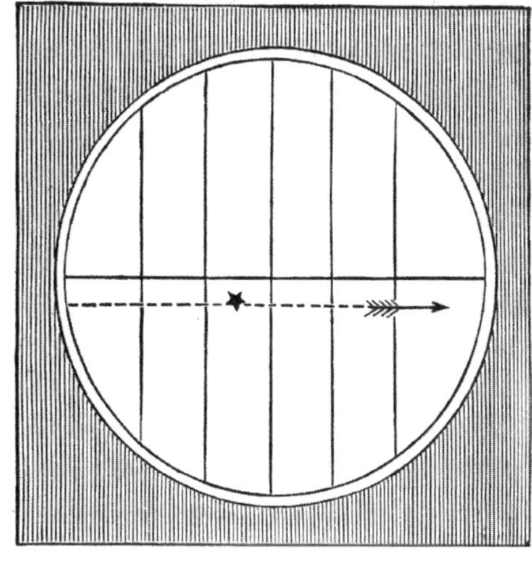

This illustration represents the view seen by an observer looking through the eyepiece of a meridian circle or transit telescope. A transit telescope does not compensate for the westward motion of the stars (caused by Earth's eastward daily rotation) as an equatorial telescope can. So a star moves steadily through the field, successively crossing the threads (or "wires"), as the arrow shows. The central thread marks the local meridian, and the observer must determine when the target crosses or "transits" the meridian. The observer would note that moment, often by pressing an electrical switch to record it on a paper tape or cylinder. The measurement of the time of meridian transit could then be converted into a measurement of the east-west location of the target among the stars. SIMON NEWCOMB, *ELEMENTS OF ASTRONOMY* (NEW YORK: AMERICAN BOOK COMPANY, 1900)

When Stebbins became director in 1922, Washburn Observatory astronomers took up the new technology of photoelectric photometry, created new research programs, and never looked back. By the summer of 1934, after several unsuccessful attempts to sell it, Washburn's Repsold meridian circle was dismantled and stored away in crates so the west wing of the observatory building could be remodeled for use as a classroom. One section of the brass tubing of the Repsold instrument was adapted to make a new special-purpose telescope in the late 1950s, but the fate of most of the Washburn meridian circle instrument remains unknown.

10

EARLY COMPUTERS OF THE HUMAN KIND

Even in the earliest days, women had important roles to play at Washburn Observatory. As was the standard of the time, the cold, dark task (accompanied by higher pay and greater recognition) of overnight observing was generally reserved for men. Exceptions were female astronomers at traditional women's schools, such as the prolific and world-famous observer Maria Mitchell, who ran the Vassar College observatory, and of course Wisconsin's own Alice Lamb. Women more typically received daytime roles, including advanced mathematical tasks, that were nevertheless as vital as observing. Without computational work, few observations could ever turn into scientific results, much less be published.

And so, from Washburn Observatory's beginning, we find "computers," which was the term used for personnel hired to carry out calculational work. The first were Elizabeth Scholfield, Emma Gattiker, and Alice J. Sanborn, who transferred data from observing notebooks into the columns of large, printed computation forms called spread sheets, then carried out whatever calculations the spread sheet was designed for. This would have included, for example, transforming the raw data from the meridian circle readings into celestial coordinates that could be plotted on a star chart and published in star catalogs. Beginning in 1881, Scholfield and Gattiker worked for Edward Holden on his project to create a catalog of stars in sections of the southern sky, based on observational data from the US Naval Observatory. Holden estimated that the labor involved in completing this task would eventually amount to between three and four million written digits, nearly all of which would have been written by the two women.

Such calculations were obviously laborious. The computers performed their mathematical operations mostly by hand, since even the most advanced calculating machines of the day were extremely rudimentary by modern standards. They also would have used aids such as special slide rules and extensive tables of logarithms and trigonometric functions. Men also

ABOVE LEFT: Elizabeth "Lizzie" Scholfield, from Wausau, Wisconsin, enrolled at the university in the fall of 1877, which means she was present on campus during the construction of the new observatory. Scholfield took Holden's astronomy course and subsequently worked for him as a "computer" on his star catalog project. Scholfield received her bachelor degree in science in June 1881. UNIVERSITY OF WISCONSIN-MADISON ARCHIVES, MADISON, WISCONSIN

ABOVE RIGHT: Emma Gattiker, of Baraboo, Wisconsin, earned her bachelor degree in science from the university in 1881. Like Scholfield, Gattiker took Holden's astronomy course and then became involved in the star catalog project. After graduation, Gattiker returned to Baraboo, where she worked as a librarian. Gattiker was active in the Sauk County Historical Society and became an authority on the local settlers who, like her parents, had come from Switzerland. UNIVERSITY OF WISCONSIN-MADISON ARCHIVES, MADISON, WISCONSIN

did such computing, of which there was plenty to go around, especially during their training or for particular research projects, as it was a task that could easily occupy a cloudy night.

While computing was a never-ending need, other observatory tasks were deemed suitable for women as well, such as the calibration of instruments. Holden entrusted Sanborn, for example, with evaluating and calibrating the meridian circle collimators, which were essential auxiliary instruments, and writing up the results for publication in volume two of the *Publications of the Washburn Observatory*. Gattiker, at Holden's direction, made use of her German language skills for the translation and publication of notes by the German optician Carl August von Steinheil regarding the making and characteristics of astronomical telescope lenses.

Alice Jane Sanborn came to Madison from Freeport, Illinois, to study at the University of Wisconsin. While working as a "computer" at Washburn Observatory, Sanborn earned a bachelor of letters in 1883 and a masters in English literature in 1886.
UNIVERSITY OF WISCONSIN–MADISON ARCHIVES, MADISON, WISCONSIN

The computers were typically students at the University of Wisconsin, so like other observatory assistants they were taking classes and participating in student life. In a March 1881 event organized by the Laurean Society, a student organization, Sanborn gave an address on the theme of ambition, which she related to the university's new 15.6-inch telescope. Sanborn would go on to win the 1883 Lewis Prize for a student-written essay, joining the august company of Lewis Prize winners Frederick Jackson Turner (in 1884) and Elizabeth Agnes Waters (in 1885), and was encouraged by Holden to study the "higher principles" of astronomy, though whether she ever did is unknown.

Working as a computer also offered monetary income and, sometimes, lodging. Gattiker, for example, worked part-time in the summer of 1883 as a student computer for a salary of twenty dollars per month (for comparison, a senior staff astronomer at the time could expect to make about $165 per month, or almost $2,000 per year). She was given lodging in a room in the Observatory Residence, which was also occupied by Holden and his family. Meanwhile, male student assistants could expect to earn part-time wages of twenty-five dollars per month plus lodging, though they were likely to be offered more rustic quarters in the observatory itself or even in the gloomy Watson Solar Observatory.

We often know relatively little about the observatory's many student assistants—sometimes not even their full names—before the middle of the twentieth century. Only if they went on to professional careers, astronomical or otherwise, are we likely to find much biographical information. Unfortunately, since professional careers were quite restricted for women in those days, the details we have about them and their lives tend to be scant. But, just as broad societal trends have changed, the field of astronomy has evolved, freeing women from narrowly acceptable roles so that everyone who wants to explore the stars can reach their full potential.

11

SELLING TIME

THE WASHBURN OBSERVATORY TIME SERVICE

Accurate information about time had long been the province of astronomers and observatories, which often used a visual signal, such as a time ball, to indicate a set time daily, to which the public could then set their own clocks. But by the 1870s, telegraph time signals, which could be transmitted over great distances, began to replace visual signals.

Banks, courtrooms, and jewelers (who sold clocks and watches) helped create a market for accurate time signals, but the most lucrative customers were the railroads, which operated on tight schedules and would pay substantial amounts to avoid accidents by keeping their station clocks synchronized to within a couple of minutes. Several observatories, including those at Harvard and Ann Arbor, established time signal services, and Chicago's Dearborn Observatory actually funded its operations for a while using income from time signals.

This was all well before Washburn Observatory came along. Astronomers everywhere found the true time by observing the meridian transits of stars by night or the Sun by day, and they used these observations to set the time-keepers, the essential observatory clocks. When equipping the new observatory, James Watson ordered one of the world's finest sidereal clocks, one that keeps time by the stars, from Anton Hohwü of Amsterdam. It served the observatory for many decades. Keeping solar time required a different clock, a standard time clock, and Watson's choice was a Seth Thomas product.

When Edward Holden became the second director of Washburn Observatory, he arrived in Madison with a plan to establish a time service in Wisconsin. Holden was well qualified for the task, having authored a report on transmitting telegraphic time signals while at the US Naval Observatory. However, he found Watson's Seth Thomas clock to be unreliable, so he convinced Governor Washburn to purchase a new standard time clock from the Howard Company of Roxbury, Massachusetts. This clock became the master clock of the observatory

Charles Morse Huffer and an unidentified woman pose with Washburn Observatory's Hohwü sidereal clock. Based on the location of the driving weight (at the top of its travel, just under the clock face), it's possible they have just finished winding the clock, which had to be done weekly. DEPARTMENT OF ASTRONOMY, UNIVERSITY OF WISCONSIN–MADISON

time service, and the observatory soon acquired a second similar Howard clock for more reliable time keeping.

Holden experimented with a number of systems for distributing correct time, the most primitive of which was in operation by September 1881. Using a telephone (said to be the first in Madison), with its microphone in the clock tower of the university's Assembly Hall (now Music Hall), astronomers up at the observatory could listen to the ticking of the tower clock and the small bell it sounded every minute, compare it with the observatory's standard time clock (probably the Howard), and then relay corrections down the hill. Holden claimed to keep the tower clock within half a minute of true Chicago time.

To transmit over telegraph, the Howard clock itself was connected to the telegraph line at the same time each day, except Sundays, a few minutes before the designated hour, generally 10:00 a.m. The clock would transmit its "ticks" every other second by means of internal cams operating a switch. At the fifty-eighth second of each minute, the tick was

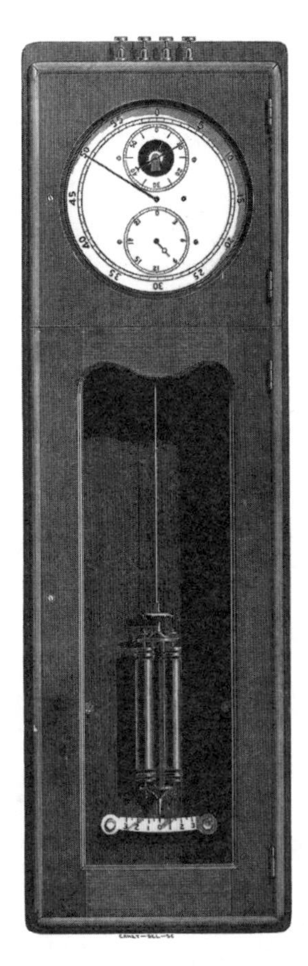

Two views of the Howard mean time clock, the principal time-keeper of the Washburn Observatory time service, which cost $375 in 1882. On the left, the cabinet doors are open; on the right, the cabinet doors are shut. The electrical terminals at the top of the cabinet were used to transmit the telegraphic time signals. The observatory's time-service assistant would find the correct time by astronomical observations, then correct this clock accordingly. This image was originally printed in *Publications of the Washburn Observatory*, Volume 2 (1884).
DEPARTMENT OF ASTRONOMY, UNIVERSITY OF WISCONSIN–MADISON

omitted, indicating that the next tick was the first of a new minute. Every fifth minute, the clock omitted the ticks for seconds fifty-two through fifty-eight. This allowed a listening telegraph operator to distinguish each minute and then the beginning of the hour.

By February 1882, Holden announced that Washburn Observatory was transmitting true Chicago time (meaning time astronomically determined in Madison and corrected to account for the longitude difference between Madison and Chicago—before the general acceptance of modern time zones). The time was transmitted daily except Sundays, although it is not clear who, if anyone, was listening to Washburn's early time signals.

Eventually Holden's sustained efforts literally paid off. By the mid-1880s, a number of railroads were purchasing time services from Washburn Observatory, including the Chicago and North Western; the Milwaukee, Lake Shore and Western; the Chicago, Milwaukee and St. Paul; the Central Iowa; and the Wisconsin Central railroad companies. At the peak, around 1887, Washburn Observatory was one of three observatories that dominated the time service business in the Midwest, along with Goodsell Observatory at Carleton College in Northfield, Minnesota, and the Washington University Observatory in St. Louis.

Former governor Washburn had left no provision for the observatory in his will, so his patronage came to a very quick end after his death in 1882. The university's funding for the observatory was essentially fixed, which meant the time service provided essential income for Wisconsin's astronomers in the early years. Railroads paid a few hundred dollars annually for the service, and although specific records are hard to find, in some years the time service income came close to matching the observatory's state-appropriated annual operating budget of $1,500. This was more than enough to fund a student assistant to tend the

View of the east wing and clock room of Washburn Observatory. The case enclosing the Hohwü sidereal clock dominates the room. The case helped insulate the clock from rapid temperature changes, thus reducing its error. Portraits of Governor Cadwallader Washburn (over the fireplace) and Director Edward Holden hang on the east wall, and a large gaslight fixture hangs from the ceiling. On the tall corner table, just below the gaslight, sits the recording chronograph, which recorded the "ticks" from the sidereal clock along with timing signals from an observer at one of the telescopes. DEPARTMENT OF ASTRONOMY, UNIVERSITY OF WISCONSIN–MADISON

clocks and telegraph operations and make the astronomical observations for time-finding. In addition, Holden sold time signals to a few watchmakers, who had to install their own telegraph lines and paid fifty dollars annually to receive the signals.

Holden also offered a system more suitable for individual offices or businesses that did not want to interpret telegraphic signals to set their clocks. The observatory provided an electrical signal over a dedicated wire to synchronize special electrically controlled clocks at locations around Madison, including the Wisconsin Supreme Court, the Western Union office, the Park Hotel, and the offices of the *Wisconsin State Journal*. These customers paid ten to twenty-five dollars per year for the electrical signal service and purchased the special clocks through Holden, who, as the manufacturer's agent, collected a nice markup. The observatory equipped the university president's office with a synchronized clock as well, and the observatory clock was used to ring the class-change bells on campus well into the 1920s.

The Washburn Observatory time service continued under George Comstock after Holden moved to Lick Observatory (where he also started up a time service). Its popularity peaked in the late 1880s, when Western Union began distributing Naval Observatory telegraphic time signals over its nationwide network. By the early 1890s, the business of observatory-run time services was diminishing as the Western Union service, which could offer substantially lower prices, flourished.

Washburn Observatory held on longer than most. The final student assistant in charge of the time service seems to have been none other than future director Joel Stebbins, who served in that role between 1900 and 1901. The service by then likely did not produce significant revenue, so the job mostly involved tending to the observatory clocks, which were essential for the astronomers, while also ringing the classroom bells.

New technology was quickly replacing telegraphic time signals anyway. In 1913, Beloit College's experimental radio station began transmitting time signals from their observatory, and by 1920, Comstock installed a wireless receiver at Washburn Observatory to pick up the radio time signals that the National Bureau of Standards had just begun to transmit.

12

KEEPING THE OBSERVATORY RUNNING

ALICE LAMB AND MILTON UPDEGRAFF

Washburn Observatory was left without a director when Edward Holden departed in late 1885. After several prominent astronomers declined the job as director, UW physics professor John Davies (1839–1900) agreed to provide nominal supervision to Holden's trusted student assistants, Alice Maxwell Lamb (1863–1952) and Milton Updegraff (1861–1938), who would keep the observatory operating, organized, and maintained until a new director could be hired in 1887.

Alice Maxwell Lamb. UNIVERSITY OF WISCONSIN–MADISON ARCHIVES, MADISON, WISCONSIN

Lamb was an accomplished young astronomer from a Madison family who defied prejudices about women's suitability for night work, in the tradition of Poland's Elisabeth Koopmann Hevelius and England's Caroline Herschel. Within a male-dominated field in the 1880s, Lamb became a skilled observer with both the 15.6-inch equatorial and meridian circle telescopes. She also had charge of the observatory time service and held the title of observatory assistant just like her male colleague Updegraff, an engineering student originally from Decorah, Iowa. Lamb seems to have worked largely at the eyepiece of the Repsold meridian circle, while Updegraff read the microscopes, though they occasionally switched roles. Lamb also observed double stars and for her work was cited by Sherburne Burnham in his double star catalog of 1900. Lamb graduated with two degrees from the university, a bachelor of letters degree in 1884 and a master of letters degree in 1885, with special honors in German. Updegraff graduated with a bachelor of civil engineering degree in 1884 and a master of sciences degree in 1886.

In the absence of a director, Lamb and Updegraff published volume four of the *Publications of the Washburn Observatory* and also produced the only significant research results of that period—meridian circle determinations of star positions needed for asteroid observations, atmospheric refraction, and several other research programs—which eventually would be published in volume six, part one of the *Publications* in 1890.

Milton Updegraff. UNIVERSITY OF WISCONSIN–MADISON ARCHIVES, MADISON, WISCONSIN

Lamb maintained a correspondence with Holden, and her letters offer many insights into life at Washburn Observatory in those days, extending well beyond astronomy. In January 1886, she wrote: "We are having a very snowy winter and it has been very amusing to observe Mr. Updegraff's persistence in appearing surprised to see me at the observatory on stormy mornings. His kind expressions of astonishment became so monotonous that I was at last obliged to reproach him with having less confidence in me than John [Doescher, the janitor]. . . . John always watches for me, and, if the drifts come more rapidly than he can shovel, helps me to find the best way through them. So for the last few days Mr. Updegraff has contrived to conceal his astonishment. Whether it has been as great as ever I do not know."

Lamb wanted to continue working in astronomy, and, as she wrote to Holden in December 1885, she was offered a teaching position running a small observatory at a school in the Dakota Territory (in present-day Yankton, South Dakota): "I refused the offer partly because I had far rather work in a true observatory than direct a toy one and partly because I felt at any rate in a measure bound to stay here until March. Of course, I should be glad to stay much longer."

As it turned out, Lamb married Updegraff in September 1887. They moved to the Argentine National Observatory in Córdoba for Updegraff's first professional appointment in what would become a long astronomical career. While at Córdoba, Lamb had charge of their observatory time service, but aside from a letter published in *Astronomische Nachrichten* in 1888, Lamb's own astronomical career did not continue. After their stint in Argentina, Updegraff worked for the University of Missouri before joining the staff of the US Naval Observatory (USNO) in Washington, DC. He served as director of the *Nautical Almanac* office for a few years before taking postings in the American West, American Samoa, and USNO's Mare Island Station. Alice and Milton raised a family and remained together the rest of their lives.

13

CONTINUITY COMES TO WASHBURN OBSERVATORY

GEORGE CARY COMSTOCK

Astronomer George Cary Comstock (1855–1934) brought for the first time stability and a coherent research program, or rather several, to Washburn Observatory as its third director—a position he would hold for more than thirty years.

George Comstock. UNIVERSITY OF WISCONSIN-MADISON ARCHIVES, MADISON, WISCONSIN

Comstock was deeply familiar with Madison, where he had been born, and with the Washburn Observatory, which he had helped build. In 1879, Comstock had returned to Madison from Ann Arbor as James Watson's assistant; then, after Watson's death in November 1880, Comstock had overseen the observatory and its ongoing construction, reporting to Edward Holden as of early 1881. Unsure of the future prospects of an astronomy career, Comstock obtained a law degree from the University of Wisconsin in 1883 while working for Holden as a student assistant, which meant running the observatory during Holden's frequent absences. Comstock worked for a few months at the US Naval Observatory but declined a permanent job offer there in early 1884; and although he considered moving to Minneapolis to open a law office, in the end he never actually practiced law.

By the time the Board of Regents was seeking a replacement for Holden, Comstock was teaching mathematics at Ohio State University. He had received offers from Holden to join the staff at Lick Observatory, but he preferred the offer from Wisconsin, and he got the job with Holden's endorsement. When Comstock was appointed in September 1887 as Washburn Observatory's associate director, the university's Board of Regents specified that he was to be under the nominal supervision of astronomer Asaph Hall of the US Naval Observatory, who had agreed to serve as consulting director. Hall was world renowned for his discovery of the moons of Mars in 1877 and was rising quickly as a major figure in US astronomy (in 1880, Governor Washburn had unsuccessfully tried to recruit Hall as director after Watson's death). As a nonresident consulting director, Hall's role at Washburn Observatory was mostly advisory. He did occasionally make the trip to Madison to work with Comstock, and they consulted by correspondence from time to time until Comstock became Washburn's director in his own right in 1889.

Asaph Hall (the third in his family to bear that name) was one of America's preeminent observational astronomers at the time he accepted the appointment as "consulting director" of Washburn Observatory. He previously studied at the University of Michigan's Detroit Observatory under Franz Brünnow, who was also James Watson's teacher, before finding a position at the US Naval Observatory in 1862, where he rose to fame. UNIVERSITY OF WISCONSIN–MADISON ARCHIVES, MADISON, WISCONSIN

Upon his return to Madison in August 1887, Comstock moved into the Observatory Residence, where he would live until 1922, chalking up the longest tenure of any tenant there. After Comstock's marriage to Esther Cecile Everett of Madison in 1894, the Observatory Residence became a busy social center for students and faculty. Comstock was also an active member of and contributor to the Madison Literary Club.

The same year Comstock returned as associate director, the Board of Regents learned that the US Naval Observatory, to which the Wisconsin observatory already had some connections, would be relocating. The regents offered the use of Washburn Observatory for any Naval Observatory work that might be disrupted during the move. Naval Observatory astronomer Stimson J. Brown accepted the invitation and arrived in Madison in late 1887 to observe with the Repsold meridian circle.

Comstock, no doubt with Hall's collaboration, kept the Repsold meridian circle busy by recruiting a succession of observers from the Naval Observatory, including eventually Albert S. Flint (1853–1923), then a junior astronomer at the Naval Observatory who arrived in Madison in 1889 when he accepted the position of assistant astronomer under Comstock. Flint's first job was to assist his Naval Observatory colleague Brown, who was using the meridian circle to improve the position measurements of cataloged stars. When Brown returned to Washington in 1890, Flint completed the work Brown had begun and then remained in Madison for the rest of his life.

Albert Flint, an astronomer at the US Naval Observatory, came to Washburn Observatory as the meridian circle specialist. UNIVERSITY OF WISCONSIN-MADISON ARCHIVES, MADISON, WISCONSIN

When Comstock became director in his own right in 1889, Hall gave up his Washburn Observatory responsibilities, but he left his mark. Influenced by famous Dutch astronomer Jacobus C. Kapteyn, Hall had suggested to Comstock that the Repsold meridian circle could be used for a program of stellar parallax measurements—an essential but difficult process that allows astronomers to measure the distance between our solar system and distant stars. Thanks to Flint and Comstock, that work would become a major activity at Washburn Observatory for the next thirty years.

Measuring the distances to other stars is essential for understanding not only the scale of our galaxy and the Sun's environment but also the physical nature of other stars, particularly their sizes and the amount of energy they radiate (known as *luminosity*). In general terms, a parallax is the apparent shift in an object's position, resulting from a change in the observer's point of view. Thanks to the Earth's motion around the Sun, astronomers do not observe the sky from a fixed point in space but from a constantly changing point of view. This causes the positions of stars to appear to shift against the background of even more distant stars. This shift is called stellar parallax and can be measured by the very small angle created when the same star is observed at two different points in time, six months apart on opposite extremes of Earth's orbit. Stellar parallax angles are so minuscule that they were detected for the first time in 1838 despite centuries of searching, thanks to improvements in telescopes and methods for measuring small angles in the sky. Only the closest stars showed parallaxes large enough to measure in those days, so constant improvement in the technologies and methods for measuring more distant stars has driven much astronomical research since the mid-nineteenth century.

The meridian circle telescope was an instrument that, in the hands of an expert like Flint, could yield data that spanned interstellar space—and in 1893, Flint, under Comstock's direction, began a program of meridian circle measurements for the determination of stellar parallax. For decades, on nearly every clear night, Flint followed the peculiar schedule of the "parallax" astronomer of those days: observing from dusk until about nine o'clock, pausing until about three in the morning, then observing until dawn. Working this way, he still managed to raise a family and participate in social life as a singer in a choral society, a sustainer of his poetry club, and an active member of his Unitarian church.

Flint's decades of stellar parallax work were very influential at just the time that photographic methods were beginning to displace the traditional meridian circle work of positional astronomy. In 1922, Kapteyn produced the most influential model of the size and shape of the Milky Way in the early twentieth century, and he acknowledged that Flint's work had provided fully two-thirds of the stellar distance information used in his model. Kapteyn's galaxy made the headlines, but Flint's decades of dedication to long hours of observing and calculation made Kapteyn's galaxy possible.

Following in Sherburne Burnham's footsteps, Comstock began a decades-long program of high-accuracy binary star measurements with the 15.6-inch telescope. Most double stars turn out to be what astronomers call binary star systems, which are systems of two or more stars orbiting each other. Comstock's observations provided the data necessary to plot the orbits of our neighboring star systems in the galaxy. Perhaps most important (and not only in his own estimation) were his studies showing that many very dim stars are in fact moving relatively rapidly across the sky (what astronomers call proper motion), suggesting that they are actually nearby despite their low brightness. This led scientists to recognize the existence of a population of dwarf stars, much dimmer than average. Another of Comstock's important research contributions was his finding that Earth's Moon has a negligible atmosphere, a conclusion he reached by observing some of his double stars in the moments just before they were occulted by the edge of the lunar disk (a lunar atmosphere should cause the spacing between the two stars to shift as they are occulted, or cut off from view, by the Moon).

Like his teacher James Watson, Comstock was a talented instrument maker, and he applied those skills to put Burnham's six-inch telescope to work by adapting an unusual instrument that allowed the telescope to look at two widely separated stars at once and accurately measure the large angle between them. This allowed him, over the course of several years, to measure two fundamental effects—atmospheric refraction and stellar aberration—that are essential for the earthbound observer to calculate the true place of a star in the sky. He also designed an ingenious device for Flint's use on the meridian circle that transformed the point image of a star into a diffraction pattern that could be more rapidly and easily measured by the observer, while also dimming brighter stars. He used a similar diffraction trick, but for a different purpose, on the 15.6-inch refractor to quantify the colors of stars.

In recognition of his scientific research accomplishments, and with Holden's assistance, Comstock was elected to the National Academy of Sciences in 1896, becoming the first University of Wisconsin faculty member to be so honored for work done entirely at the university. (Watson had been elected to the Academy in 1868 before he came to Washburn Observatory, and Holden was elected in 1885 for work done in part prior to his Washburn tenure.) In 1899, Comstock helped found the Astronomical and Astrophysical Society of America, which later would become the American Astronomical Society. He used his prior legal training to draw up the organization's first constitution and bylaws and served as its first secretary. The university turned to him in 1906 to put advanced work and research on a firm footing by organizing the university's graduate school, which he led for many years, in addition to running the observatory.

Comstock retired from the university in 1922, but his devotion to astronomy wasn't over. From 1925 to 1928 he served as president of the American Astronomical Society. After his term ended, Comstock and his wife set out on a tour around the world, visiting major observatories along the way. His life's work was generally in the mold of an older astronomy, which he typically called "astronomy of precision." But by the time Comstock died in 1934, he had witnessed the astronomy of his youth take its place beside what we now call astrophysics and had made important contributions to that transformation itself.

14

FILAR MICROMETERS

THE ESSENTIAL TOOL OF THE ASTRONOMY OF PRECISION

George Comstock characterized the research of Washburn Observatory during his career as "the older astronomy of precision" to distinguish it from the growing activity in the late nineteenth century that we now call astrophysics. One of the big differences between the old astronomy of precision and the new astronomy of astrophysics was the importance of the human eye as the preeminent detector of light. In the old model of astronomy, before photography dominated imaging and photoelectric cells measured brightness, the astronomer's eye did the work.

To measure the positions of stars and other celestial objects, astronomers used an indispensable instrument called a filar micrometer. Filar refers to the fine filaments or wires (often made from a strand of silk from a spider's web or a silkworm cocoon) used to create a series of visible reference lines, known as a reticle, in the eyepiece of the micrometer. The lines of the reticle could be vertical or horizontal or both. Typically, all but one of the wires were in fixed positions across the visual field, while the final wire was attached to a frame that could be moved across the visual field by turning a finely threaded screw. The spacing between the fixed and movable wires could be adjusted and precisely calibrated, allowing the astronomer to convert the number of turns of the screw into an angular distance on the sky. In addition, many micrometers could rotate about the optical axis of the telescope to allow measurements of position angle, such as the orientation of the line connecting the two stars of a binary pair. A good filar micrometer was a delicate and expensive accessory to the astronomical telescope.

Washburn Observatory's first filar micrometer, purchased by former governor Washburn himself in 1878, was made by Alvan Clark and Sons, who also made the 15.6-inch telescope for which the micrometer was an accessory. The Clark micrometer was first put to work

George Comstock used the Clark micrometer, shown here mounted at the "eye end" of the Washburn 15.6-inch refractor, for decades of double star measurements. Comstock also used the micrometer on the Burnham six-inch telescope for his refraction and aberration project. UNIVERSITY OF WISCONSIN-MADISON ARCHIVES, MADISON, WISCONSIN

by Sherburne Burnham in the summer of 1881 to measure the spacing and position angle of binary stars, the prolific discoveries of which had made his fame. This kind of work, tracking star systems that can have orbital periods of decades or centuries, was continued for decades by Comstock. By measuring the spacing and position angle of the components of a binary star year after year and decade after decade, astronomers could deduce the orbits of the pairs of stars, revealing among other things the relative masses of the two stars. If the distance from Earth to the binary pair could be measured (generally by parallax), then the actual

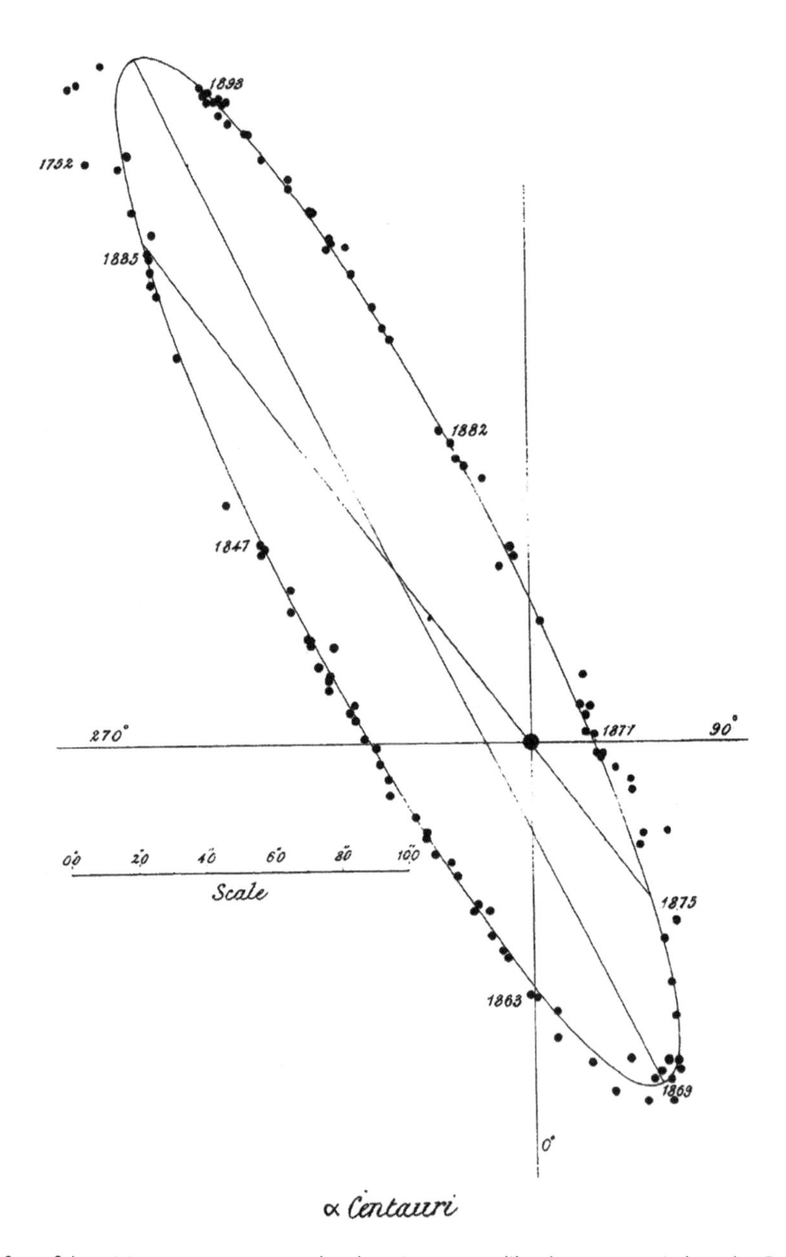

Decades of careful position measurements taken by micrometer, like the ones carried out by Comstock for many years, could help astronomers determine the orbits of binary stars. This illustration shows the visible orbit of the Alpha Centauri star system, our Sun's nearest neighbors, derived by astronomer Thomas Jefferson Jackson See in 1893 based on observations by several astronomers. Notice the scatter from observational error in the individual observations (dots). All other binary star orbits, which generally were much more distant, would be far more difficult for astronomers to determine. *MONTHLY NOTICES OF THE ROYAL ASTRONOMICAL SOCIETY* 54 (1893)

masses of the stars could be found. Herein lies one of the vital insights from the astronomy of precision that informed the astrophysics of stars: our first determinations of the fundamental property of mass for stars beyond our Sun. Comstock's decades of measurements of binary stars were considered particularly valuable because of the continuity of instrument and observer, and thus the presumed consistency of systematic errors.

The motion of a particular star or star system across the sky relative to other stars is called proper motion. Using his filar micrometer, Comstock applied his abundant double star observations to compare the proper motion of the double stars to nearby reference stars in order to accumulate measurements of a statistically significant number of proper motions of fairly faint stars, those averaging magnitude 9.5—or about twenty-five times dimmer than what can be seen by the human eye. (Astronomers traditionally rank the apparent brightness of stars by their magnitude. The higher the magnitude number, the dimmer the star. For example, magnitude 1 represents the brightest stars, like Vega, and magnitude 6 denotes the dimmest stars visible with the human eye.) From this wealth of positional data, Comstock concluded that the average proper motion of faint stars was "considerably greater" than would be expected if their faintness was generally because of their greater distances compared to brighter stars, thus "the average faint star is nearer than has been supposed."

This was the first evidence of the stellar class that astronomers began to call dwarf stars, which is to say the vast majority of stars in the universe, which are orders of magnitude fainter than the more prominent but much rarer stars. This provided some of the earliest evidence of the vast range that exists in stellar luminosities. Here again, the astronomy of precision provided a fundamental fact about stellar populations, which remains a very active field of research in modern stellar astrophysics.

Washburn Observatory acquired its last filar micrometer in 1897, an advanced transit micrometer from Repsold, as an accessory to the Repsold meridian circle, then in nearly constant use by Albert Flint for his stellar parallax measurements. This unusual micrometer's entire eyepiece and reticle could be moved east to west along a threaded shaft as the observer rotated a calibrated knob to follow the target star across the field. Electrical contacts activated by the rotation of the knob signaled every tenth of a revolution, as well as the moment when the center of the moving reticle crossed the meridian. The time signals were recorded on paper by the chronograph for analysis after the observing session. While Flint found this new method difficult to learn at first, he was, after "considerable practice," able to make measurements with the Repsold filar micrometer that were as precise as the old method (a probable error of three-hundredths of an arc second!), but with the advantage that the new micrometer reduced the variations in the observer's personal equation (or the correction that was needed to account for an observer's idiosyncratic errors), which then also reduced the differences between personal equations of different observers and improved the precision of the results.

Astronomers of the Comstock-and-Flint generation, just like the couple of generations before them, spent countless hours throughout their long careers pointing their telescopes from star to star. At each star, with their eye at the eyepiece and their hands on the micrometer controls, they made a series of measurements, often while calling out their readings to an assistant recording them in a notebook, before moving on to the next target. Their accumulated work occupies hundreds of pages of the *Publications of the Washburn Observatory*—seemingly endless tables of position measurements of one kind or another.

The enormity of this work can be seen in Flint's parallax observations, the bulk of which were recorded by the transit micrometer as he rotated a knob to track the star. The micrometer's electrical signals were marked on the paper of the rotating chronograph

This is the sophisticated transit micrometer made for Washburn Observatory by Repsold in 1897. It is still attached to the tailpiece of the meridian circle itself, dated 1882. The electrical signals it automatically generated as the observer tracked a star across the field of view helped improve the results of meridian circle measurements. JAMES LATTIS

drum, and those marks had to be measured the next day with a ruler, either by Flint himself or his computing assistant, Edna M. Hill, and averaged to derive a transit time. In his second series of parallax measurements, from 1898 to 1905, Flint recorded data for 366 stars (124 of which were parallax stars) that amounted to 29,100 observed star transits. For each of these star transits, twenty chronograph signals needed to be measured.

Edna M. Hill, an observatory assistant and computer, worked with Albert Flint from about 1914 to 1916. UNIVERSITY OF WISCONSIN–MADISON ARCHIVES, MADISON, WISCONSIN

By the early twentieth century, photography had transformed how positional astronomy was done. The new generation of observers collected hundreds of star images on a single photographic plate, and a new generation of human "computers" working during the day measured the relative positions of those stars on the plate. If there was any doubt of a given measurement, it could be repeated, unlike the fleeting transit of a star through the field of the meridian circle. The photographic plate was a record of a part of the sky at that moment, which could be filed away and reexamined as needed. Photography rendered the filar micrometer and the transit telescope obsolete for most of their important functions. Astronomical science, as well as the skills and demands of the astronomers' discipline, would never be the same.

15

WISCONSIN INNOVATION IN THE OLD ASTRONOMY

Despite being older, George Comstock's astronomy of precision was no less important in the work of the new astrophysics. For example, the actual size and energy output of a star cannot be determined from its apparent brightness unless its distance is known, and measuring astronomical distances by parallax is a classic function of the astronomy of precision.

To accurately measure parallax motions (or nearly any celestial position data), an Earth-based observer must correct for various complicating factors. Two of the most important factors are the effects of atmospheric refraction and the effects of Earth's orbital motion, which astronomers call the aberration of starlight, both of which shift the apparent position of a star away from its true position. In 1887, Comstock, who had only recently become the observatory's associate director, had the idea to combine a novel optical device called a Loewy prism (after its inventor, French astronomer Maurice Loewy) with a telescope to create an instrument that could measure both of those effects in a new way. In effect, Loewy's device allows a telescope to look in two widely separated directions at once, giving it, in a sense, a very wide field of view. Two widely separated stars would appear together in the telescope field, and the actual angular distance between them would be the separation measured by the filar micrometer in the eyepiece, plus exactly one hundred twenty degrees introduced by the Loewy device. Normally the filar micrometer could only measure a tiny patch of celestial real estate—but with the Loewy device, the micrometer's measuring precision suddenly was applicable to very wide angles across the sky.

Atmospheric refraction—or the bending of starlight away from the straight path it would follow in the absence of the atmosphere—causes a star to appear higher above the horizon than it really is; this effect is much stronger near the horizon and decreases to zero at the zenith. Refraction near the horizon is so strong that the setting Sun is actually already below the horizon when we see it setting, the rays of sunlight being bent toward the observer by refraction of the thick atmosphere. By using the Loewy device to measure the apparent angular distance between two widely separated stars, one high in the sky and the other nearer the horizon, and comparing those results with their true separation, Comstock could measure the effect of atmospheric refraction. Previously, the best determinations of atmospheric refraction had been done at Pulkovo Observatory in Russia and at the Royal Observatory near Cape Town, South Africa. Comstock hoped to determine whether the effect might be different in North America, where the measurement had not been done, and use his new method to refine previous measurements.

Aberration of starlight results from Earth's annual orbital motion and produces a periodic shifting of the apparent positions of stars. To aim a telescope accurately, the astronomer must tilt the telescope very slightly in the direction of Earth's motion, just as when walking with an umbrella, we must tip the umbrella slightly in the direction of our own motion for it to effectively block the rain. The apparent displacement of the star in the sky varies as the direction of Earth's motion changes in the course of the year. This effect had been measured with increasing precision since 1727 (it was, in fact, the first observational evidence of Earth's orbital motion), but Comstock's new method would test, and perhaps improve, those previous experiments.

In both cases Comstock was making a differential measurement between two stars, which provided some advantages compared to measuring the individual star positions with respect to zenith and horizon, as previous measurements of both effects had done.

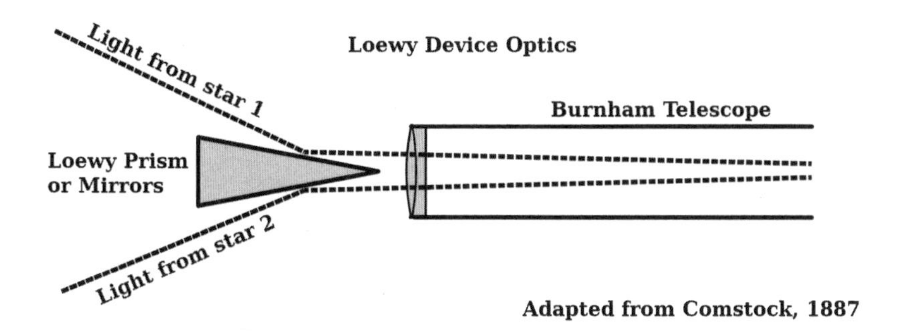

This adaptation of George Comstock's original illustration shows a cross-section of the Loewy device to explain how the device works. Two reflective surfaces (the longer sides of the triangle on the left) reflect light rays from widely separated stars into the objective lens of the telescope on the right. Though far apart in the sky, the two stars appear close enough in the field of the telescope that their separation can be accurately measured with a micrometer in order to reveal the effects of atmospheric refraction and the aberration of starlight caused by Earth's orbital motion. ADAPTED FROM GEORGE COMSTOCK, *SIDEREAL MESSENGER* 6 (1887)

Both types of measurements using the Loewy device required a wide field of view over the sky—far larger than is normally visible through the narrow slit of an observatory dome. To accommodate this, Comstock decided to radically rebuild the dome of the Student Observatory. He cut the dome into two half-hemispheres, or semidomes, and removed the one containing the dome slit. In its place he built another semidome large enough to fit around the remaining half so that the inner semidome could nest within the new outer semidome, which ran on its own track around the dome base. Thus, he could open up the observatory to see one half of the sky at once.

In 1887, about the time he was considering the dome modifications, Comstock published a short article that described how the Loewy device could be used and invited someone to

Comstock rebuilt the Student Observatory dome as a pair of nesting hemispheres that could be rotated to fully reveal half of the sky. Pictured here in the dome is the Burnham telescope outfitted on the front end (on the right) with the Loewy device and on the viewing end with the Clark micrometer. The Burnham telescope at this time still had its original Clark mounting and wooden tube. DEPARTMENT OF ASTRONOMY, UNIVERSITY OF WISCONSIN–MADISON

build one. His article drew the attention of instrument maker John A. Brashear, of Pittsburgh, Pennsylvania, who offered to help, and Comstock applied for and (fittingly) received funding from the Watson Fund of the National Academy of Sciences to pay for the device's fabrication. The process, however, proved more difficult than either Comstock or Brashear had expected, because each face of the wedge-shaped prism had to be made optically flat, but the thermal properties of the glass, which cooled more rapidly at the thinner end during the optical polishing, proved so hard to control that Brashear gave up. Finishing the device required that they make substantial design changes, primarily by replacing the surfaces of Loewy's prism design with flat mirrors. Instrument maker George Saegmuller, of Fauth and Co. in Washington, DC, made a special coupling to hold Brashear's Loewy device on the end of the Burnham telescope, which had to be shipped to the Fauth shop to be fitted. With the Loewy device mounted on the front of the telescope and the Clark micrometer in the eyepiece, Comstock was ready to begin his observing program in September 1890.

By the summer of 1892, Comstock had completed his observational work, but it took another two years of computing (done manually, of course) to produce his results. In the end, he published a new, more convenient way for astronomers to calculate the atmospheric refraction effect, as well as one of the best determinations of the aberration effect that had been made up to that time. Comstock's use of the Loewy device was only one of his many innovative instrument projects, all of which involved making telescopes produce better measurements, often by optical tricks like the Loewy device and diffraction effects.

Comstock's aberration-refraction project marked the end of the research career for Burnham's six-inch refractor. It would remain in place at least until the late 1920s in the Student Observatory, which, despite the need for the special semidome having passed, wouldn't be returned to its more conventional configuration with a narrow-slit opening until about 1930. The Burnham refractor has enjoyed an active retirement, as it's been used as an instructional tool by many generations of astronomy students to the present day.

16

DIARY OF AN ASTRONOMY STUDENT

SIDNEY TOWNLEY

In the fall semester of 1886, Sidney Dean Townley (1867–1946), from Waukesha, Wisconsin, enrolled as a freshman at the University of Wisconsin. During his college years, Townley would become interested in astronomy, earn a distinguished scholarship, work as an assistant astronomer, and receive a master's degree, before going to California and then Michigan for his doctoral studies and ending his career as a professor at Stanford University.

Sidney Townley, in a photograph from the Washburn Observatory photo album.
UNIVERSITY OF WISCONSIN–MADISON ARCHIVES, MADISON, WISCONSIN

While a student in Madison, Townley kept a diary that provides a unique and fascinating glimpse into student and observatory life at a time when the university was rapidly growing. Townley fully immersed himself in campus endeavors, serving as editor of the junior yearbook, class president, president of the student literary society, and an avid baseball player, manager, and umpire.

Townley's diary recounts adventures of campus life, including an amphibious raid launched from a borrowed rowboat into the university orchard that netted three bushels of apples, the hazing trials in which sophomores "sentenced" freshmen to sing or dance, and riotous Halloween rampages through town.

The northern slope of Observatory Hill, as seen from Lake Mendota. In the foreground is the university orchard. From left to right, along the crest of the hill, we can see the barn where Sidney Townley lodged his horse, the Student Observatory, and the main observatory building (the roofline of the Observatory Residence can be seen rising behind the barn, as well). Note the absence of the balcony outside the dome of the main building, which wasn't added until about 1927. UNIVERSITY OF WISCONSIN–MADISON ARCHIVES, MADISON, WISCONSIN

And he describes elaborate pranks, such as a February dance party outing in Middleton during which undisclosed persons surreptitiously hitched the horses and sleigh and drove them back to Madison, stranding the couples who, in those straightlaced days, then had to pass the rest of the night awkwardly together at the rural dance hall. Townley's account mentions the origins of the Badger yearbook and the *Daily Cardinal* student newspaper, as well as the excitement over the new Science Hall. His diary bears witness to the formation of the university we know today.

Townley's interest in astronomy was sparked in January 1888 by Richard Proctor, the famous speaker, author, and astronomy popularizer who was brought to campus as part of the *Peoples Lecture Course*. The exciting lecture was "illustrated throughout by pictures thrown on a screen with a magic lantern." That spring semester, Townley took the introductory astronomy course taught by new observatory director George Comstock. By the beginning of Townley's junior year the following fall, he was working at the observatory as a night assistant, helping Comstock with double star observations, for which he received twenty cents per hour. The job also included lodging in Washburn Observatory itself at a cost of four dollars per month, and the great privilege of round-the-clock access to the observatory's Woodman Astronomical Library.

Upon his hiring, Townley became part of the observatory "family," which was known for looking after its students. For example, when Townley returned to Madison for the fall semester of his senior year, the train arrived late at night and he had nowhere to stay. He went up to the observatory, found an open window, and spent the night on the sofa in the clock room. He awoke in the morning to find a note from Alice Brown, wife of meridian circle astronomer Stimson Brown, inviting Townley to the Observatory Residence for breakfast. The Browns then kept him fed until he could arrange room and board in town.

Upon graduation, Townley's academic performance earned him one of the new university fellowships for graduate study, worth four hundred dollars per year for up to two years. Townley chose to continue his study of astronomy with Comstock. Early in his first fellowship year, he went home to Waukesha to get the family horse, Kit, and bring her to Madison. His account of their two-day rides from Waukesha and back provides a fascinating portrait of a now-vanished rural Wisconsin, as they helped themselves to apples and corn along the way, stopping overnight in the very German settlement of Jefferson and returning via Fort Atkinson the following spring. He lodged Kit in the observatory barn that fall so that he could take her out for occasional rides in the country while he was learning from Comstock the traditional visual observing methods of double and variable stars, while also serving as umpire for undergraduate baseball games.

Ultimately, Townley would publish useful accounts of the Washburn Observatory time service, with which he was intimately familiar, as well as his thesis research on measurements of long-period variable stars, which would appear in the *Publications of the Washburn Observatory*. Although he hoped to move on to Lick Observatory in 1891, no funding was available that year to his support his studies. Though he was tempted to pursue further astronomical study at Harvard, the following year Townley chose to accept, with Comstock's recommendation, the first Phoebe Hearst Fellowship at the new Lick Observatory, under director Edward Holden. Not coincidentally, the position brought him closer to a young woman in Oregon whom he had met the previous summer and would eventually marry.

17

TRAINING ASTRONOMERS

THE PERSONAL EQUATION MACHINE

Meridian telescopes, which developed rapidly in the nineteenth century, generally work in combination with a clock. The astronomer would observe the transit of a star through the eyepiece of the telescope, which contained a line marking the meridian. The observer's job was to determine the instant when the star crossed the meridian line and record the time of that event by the clock. The time difference between two such meridian transits allowed observers to measure the east-west angular distance between two objects. These could be two stars or perhaps a star and an orbiting object, such as a comet, planet, or asteroid.

Washburn Observatory's personal equation machine. JAMES LATTIS

Portrait of Joel Stebbins from the Washburn Observatory photo album when he was a graduate student under George Comstock in academic year 1900–1901. The next year would find Stebbins at Lick Observatory, where he would receive a PhD in 1903. UNIVERSITY OF WISCONSIN–MADISON ARCHIVES, MADISON, WISCONSIN

Astronomers quickly discovered that even seasoned observers introduced systematic errors into this measurement—for example, some observers slightly anticipated the true moment of transit, while others lagged behind it. The error was idiosyncratic, but a given observer's tendencies were generally predictable and could be corrected through the computing process once their tendencies were known. This correction was known as the personal equation.

By the late 1800s, observers commonly registered the time of transit by pressing a button or key that would send an electrical signal to a chronograph, which would record periodic timing marks, like visible ticks of the clock, on chart paper on a slowly rotating drum or later on paper tape, along with the observer's signal. The time of the observer's mark could be interpolated from its location among the clock's regular ticks. This did not eliminate the personal equation, but it did provide a permanent record of the measurement.

Meridian circle astronomer John R. Eastman of the US Naval Observatory invented a personal equation machine to allow astronomers to measure their own personal equation during the day, thereby sparing valuable telescope time at night. The machine used a mineral oil lamp to illuminate a small aperture, creating an artificial star projected on a ground glass plate with lines similar to those seen in a transit telescope field. The observer would look through a tube, simulating the telescope, as the artificial star crossed the "telescope" field, moved by a wind-up clockwork mechanism. An electrical contact built into the machine would automatically send a signal at the true moment of transit, recorded by a mark on the paper tape of the chronograph. Meanwhile the observer would press a key at the perceived moment of transit to make another mark on the tape. The time difference between the true mark and the observer's mark, measured over many trial runs, would be used to find the observer's personal equation.

In 1898, George Comstock purchased an Eastman machine, made by the Kahler instrument firm in Washington, DC, which meant it was still a relatively new tool at the Washburn Observatory when Joel Stebbins arrived in Madison in 1900 for his first year of graduate study. Stebbins recorded his exercises with the personal equation machine in a log book beginning March 24, 1901, showing the data from 187 trials.

Stebbins spent one year studying in Madison. In addition to measuring his personal equation, his student training at Washburn included learning to use the Bamberg "broken" transit instrument in the Student Observatory and probably the much larger Repsold meridian circle as well. When Stebbins eventually returned as Washburn's director in 1922, he had no use for these instruments of the old astronomy, which were irrelevant to his research in photoelectric photometry. Nonetheless, as Stebbins wrote to his predecessor and mentor Comstock soon after returning to Madison, "I was interested to see how I could manipulate the Bamberg transit after some twenty years, and I give the results of my first and second nights. . . . These are not so bad." For Stebbins, the transit work of his youth had become little more than a nostalgic curiosity in the new age of astrophysics.

Scan the QR code* to watch a video of Washburn Observatory's original personal equation machine in action.

*Or visit go.wisc.edu/personalequation

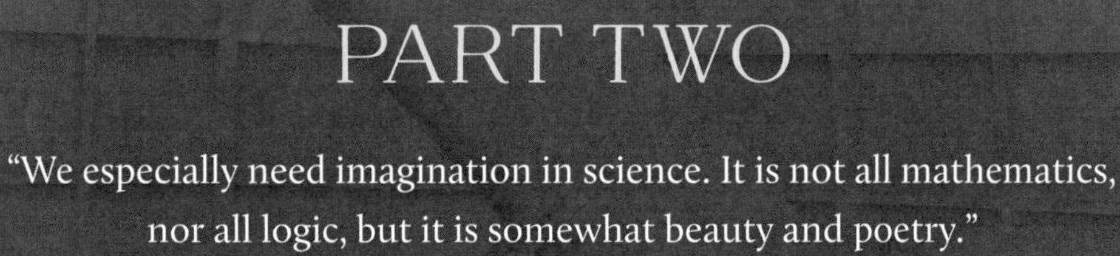

PART TWO

"We especially need imagination in science. It is not all mathematics, nor all logic, but it is somewhat beauty and poetry."

From an 1871 journal entry by Maria Mitchell, the first professional female astronomer in the United States and the first woman elected to the American Academy of Arts and Sciences

18

PIONEER OF ASTROPHYSICS

JOEL STEBBINS

A major turning point for Washburn Observatory came when Joel Stebbins (1878–1966) succeeded George Comstock as director in 1922—a transition that was more than twenty years in the making. At a time when many nineteenth-century observatories were lapsing into obsolescence, Stebbins would transform Washburn Observatory from a stronghold of the astronomy of precision into a vehicle in the vanguard of astrophysics. It took him about a year.

Albert Flint had retired in the summer of 1920, and the Repsold meridian circle remained untouched after he left. Comstock had been planning his own retirement as well, with his former student Stebbins as his arranged successor. Though Comstock had effectively ended his own observing activities, he agreed to continue serving as the observatory's temporary administrative director so that Stebbins could have time to transition before leaving his position at the University of Illinois in Urbana.

Stebbins spent about a year visiting Madison on frequent trips, appointed by the university as a nonresident professor, as he began preparing Wisconsin's observatory for radically new directions. He officially assumed Washburn's directorship on July 1, 1922, shortly after moving into the Observatory Residence with his wife, May Prentiss, and their two children, Robert and Isabelle, who were fifteen and twelve years old respectively. Both Joel and May had graduated from the University of Nebraska, and May had worked in Washington, DC, as a cataloger for the Library of Congress before their marriage. By 1922, May had become an activist and organizer for international peace movements. She and her children were the first of what would become Madison's Bahá'í Faith community, to which she had converted while they lived in Urbana. (Joel did not share May's devotion to the faith.)

Stebbins already had solid Wisconsin connections when they arrived. His astronomy teacher at the University of Nebraska, Goodwin D. Swezey, was originally from Beloit,

Wisconsin, and had spent several summers at Washburn developing his astronomical skills as an unpaid observatory assistant under Comstock. When the young Stebbins sought graduate studies, Swezey recommended him to Comstock, and Stebbins studied with Comstock at Washburn during the 1900–1901 academic year. Comstock then recommended Stebbins for a fellowship at the University of California's Lick Observatory. Stebbins graduated from the University of California with a PhD in 1903 and was immediately hired by the University of Illinois as observatory director, where he worked until his return to Madison.

During his nearly twenty years at the observatory in Urbana, Stebbins established himself as an innovator and researcher in the field of photometry—which allows scientists to measure the brightness of stars—by inventing light-sensing instruments far more sensitive and precise than the human eye or photographic plate. In the early days at Urbana, Stebbins

This portrait of Joel Stebbins as director of Washburn Observatory was probably taken in the 1930s. Stebbins sits at his desk in the observatory library, which also served as the director's office.
UNIVERSITY OF WISCONSIN–MADISON ARCHIVES, MADISON, WISCONSIN

worked with a primitive optical photometer in which the observer's eye was still the judge, the "detector" so to speak, in making a measurement. He began collaborating with University of Illinois physicists—at first F. C. Brown and most notably Jakob Kunz, who would remain a longtime collaborator—to develop astronomical photometers that eliminated the imprecise human eye from the process of measurement. Stebbins and Brown developed an electrical photometer using the properties of the photoconductive metal selenium. The selenium detector was a great advance over the human eye, and Stebbins and Brown developed instruments that could measure previously invisible fluctuations in the brightness of variable stars, work that rapidly advanced Stebbins's research and propelled his election to the National Academy of Sciences in 1920. Stebbins's collaboration with Kunz, an expert on the photoelectric effect, began in 1911. Using photoelectric detectors from Kunz's lab at the University of Illinois, Stebbins and Kunz were able to produce photometers that could significantly outperform the selenium photometer, which was abandoned by 1913.

As Stebbins began his transition to Washburn Observatory, he could not bring with him the photoelectric photometer that belonged to the University of Illinois. So, from the autumn of 1921 to the summer of 1922, the University of Wisconsin funded the construction of a new one in preparation for the new director's arrival. Stebbins took the opportunity to introduce some improvements to the new photometer. Steady technical improvements by Stebbins and his collaborators over the next ten years would establish the photoelectric photometer as one of the pillars of modern astrophysics. By the summer of 1923, Stebbins had his first Wisconsin-built photoelectric photometer working, swiftly pivoting Washburn Observatory away from the astronomy of precision in its traditional practice.

While photography replaced the human eye in most research observatories, Washburn Observatory skipped the ascension of photography and jumped directly from human observers, like Comstock and Flint, to Stebbins's photoelectric photometers containing Kunz's light detectors, which are the distant ancestors of modern electronic light-sensing devices. The new technology would help to drive many of the advances that happened at Washburn under Stebbins, from binary star research and solar eclipse expeditions to developing a way to measure the effective color of a star, which allowed Stebbins to be the first to calculate the effects of interstellar matter and estimate the approximate true size of our Milky Way Galaxy.

In addition to Kunz, Stebbins actively recruited the talents of other scientists to help accelerate Washburn's innovations. Most notably these included Charles Morse Huffer, who served as Stebbins's second astronomer, and Albert Whitford, a student assistant who later would become Stebbins's handpicked successor as Washburn's director.

Joel Stebbins and his granddaughter Joanne Stebbins unveil a portrait at the Pine Bluff Observatory dedication ceremonies on June 30, 1958. The portrait was painted in 1948 by university art professor Roland Stebbins (no relation). For years, the two Professors Stebbins had exchanged stories of confused art students inquiring at the observatory, and astronomy students looking for their teacher in the art department, which led the artist to paint the astronomer. The painting remained in the artist's possession until it was purchased for the Astronomy Department in 1958 with personal contributions from Albert Whitford, Morse Huffer, Arthur Code, and others. UNIVERSITY OF WISCONSIN–MADISON ARCHIVES, MADISON, WISCONSIN

The gregarious Stebbins was well known far beyond Wisconsin. He traveled frequently to other research laboratories, conferences, and speaking engagements, and he served as secretary of the American Astronomical Society from 1918 to 1927 and president from 1940 to 1943. Of the many awards and honors Stebbins received during his career, perhaps the most significant came in 1931 when he was named a Carnegie Institution Research Associate, which granted him observing privileges at California's Mount Wilson Observatory northeast of Los Angeles. Built and operated by Carnegie, Mount Wilson Observatory hosted what was then the world's largest telescope—George Ellery Hale's 100-inch Hooker reflector. For more than a decade, Stebbins traveled to Mount Wilson every summer, where he and his Wisconsin associates, especially Whitford, became regular guest observers and familiar figures. The combination of Wisconsin photometers with the large Mount Wilson telescopes in the clear California skies proved to be very productive. Hale enlisted and funded both Stebbins and Whitford to develop photoelectric photometers intended for the Mount Palomar 200-inch telescope, then under construction.

By the 1940s, the years on the Madison campus had left the old Washburn Observatory with an outdated telescope in the middle of a smoke- and light-polluted city. Stebbins decided it was time to look for a new location and funding for a new observatory (or "country telescope," as Stebbins liked to call it) that would give Wisconsin astronomers updated research capabilities. Before he retired in 1948, the seventieth year for both Stebbins and the observatory itself, he began the work of transitioning Washburn Observatory from a research institution, which reported directly to the university president, into an actual Department of Astronomy, under the dean of Letters and Science. The new observatory—Pine Bluff Observatory, west of Madison near Cross Plains—would eventually be completed under Whitford's directorship. And in 1958, Stebbins would return to Wisconsin from his retirement home in California, near Lick Observatory, to give the country telescope's dedicatory address.

SPOTLIGHT

HOSTING THE NATION'S ASTRONOMERS

By the end of the nineteenth century, astronomy had gained recognition as a discipline, but it had not yet fully matured into a recognized profession. Astronomers typically worked for academic institutions where they were expected to spend most of their time teaching, and only a few institutions existed that emphasized astronomical research over instruction. At the same time, scientific disciplines were rapidly becoming more specialized, and professional societies were springing up to help scientists share resources, methods, and political leverage.

In 1899, a group of astronomers gathered at Yerkes Observatory in Williams Bay, Wisconsin, to found the American Astronomical Society—or the Astronomical and Astrophysical Society of America, as it was originally called. Their aim was to professionalize the field of astronomy and, in part, to lend legitimacy to the emerging field of what we now call astrophysics. Founders included George Ellery Hale and Washburn director George Comstock.

Twenty-eight years later, in early September 1927, society members gathered at Washburn Observatory for the first time, thanks to the coincidence that two society officers lived in Madison: Comstock, who had retired from Washburn in 1922, was serving his last year as the society's president, and Washburn's new director, Joel Stebbins, was the organization's outgoing secretary. This was the thirty-eighth meeting of the growing astronomical society, and forty-eight members attended, about one-tenth of the entire membership. One of the visitors was astronomer Issei Yamamoto, who six months later would host the Comstocks at his observatory in Kyoto, Japan (and who would fondly remember Tycho, the Stebbins family dog).

The visiting members were lodged in fraternity houses along the shore of Lake Mendota, which were available because in those days the fall semester started considerably later than is now typical. The scientific sessions were held in Bascom Hall, but May and Joel Stebbins

Astronomers at the 1927 meeting of the American Astronomical Society gather outside of Washburn Observatory for a group photo featuring one of the Burnham relics, the original wooden tube of his six-inch Clark refractor (by then long replaced by a steel tube) in the foreground. Presumably that is the Stebbins's dog, Tycho, who seems interested in something inside the tube. Morse Huffer (in the round glasses) and Jakob Kunz (in the bow tie) are in the top row, and Joel Stebbins and George Comstock are in the second row from top (to the left and right of Kunz, respectively). DEPARTMENT OF ASTRONOMY, UNIVERSITY OF WISCONSIN–MADISON

hosted the opening reception at the Observatory Residence. And, of course, there was a tour of Washburn Observatory, where the famous and recently remounted Burnham telescope and its separate parts, the original wooden tube and Clark mounting, were relics of special interest. (The legendary Burnham himself had died in 1921.) Also included in the conference was a "delightful drive" to the north shore of Lake Mendota for a reception at the Comstocks' summer cottage.

The American Astronomical Society would meet in Madison several more times, in 1946 (two years before Stebbins retired), 1958 (during which Stebbins returned to give Pine Bluff Observatory's dedicatory address), 1978, and 1996.

19

THE ESSENTIAL ABSENCE

JAKOB KUNZ

Physicist Jakob Kunz (1874–1938) was never on the staff of Washburn Observatory, though he was a familiar, indeed essential, figure and visitor there. As Joel Stebbins's longtime collaborator, Kunz made his most important contribution to Washburn not in Madison but in his laboratory at the University of Illinois, where he made the unique photoelectric detector tubes, generally known as Kunz tubes, that Wisconsin's astronomers used in their astronomical research from 1922 until after World War II.

Kunz had originally made the light-sensitive photoelectric tubes partly as experiments in recording sound on film for "talking pictures." In 1911, Kunz suggested to Stebbins that the tubes might be useful in the measurements of variable stars. From that point on, Stebbins and Kunz became close collaborators, and they and their families became close personal friends. After World War I, Kunz and his wife, Anna Bolliger, became involved in the international peace movement and the Bahá'í Faith, both of which they introduced to Stebbins's wife, May Prentiss. Far from a stereotypical rationalist, Kunz sought to experience nature at a personal level. In 1923, on his way home from their solar eclipse expedition, Kunz stopped over in the Grand Canyon, later writing to Stebbins, "I went down the Bright Angel Trail and wanted to walk (after having seen the river) over in the moonshine to Hermit's Trail. The moon set about 9:30 and I lost the path. So I slept on a stone. It was the most wonderful night I ever had. All words for the Grand Canyon are too insignificant."

Originally from Switzerland, Kunz earned his PhD in Zurich, then studied at the Cavendish Laboratory at Cambridge University in England before immigrating to the United States, where he soon was hired as a professor of physics by the University of Illinois. His European education prepared him well to teach the new theories (then not well understood in the United States) of relativity and quantum mechanics, for which the photoelectric effect was a central problem.

The Kunz tubes were the most sensitive photoelectric cells available before the invention of the photomultiplier tube, which would not come into broad use until the 1950s. The construction of the highly sensitive tubes was, to some extent, more art than science, and also a matter of luck. Stebbins understood this intimately, as he himself had assisted Kunz in the complex fabrication of the tubes and even after moving to Madison would occasionally return to Urbana to help out in Kunz's lab. For other purposes less exacting than stellar astronomy, Kunz made tubes out of Pyrex and other types of glass, which he was able to do in his own laboratory. But the most sensitive tubes were made from fused quartz. When this proved too difficult for Kunz's laboratory, Stebbins negotiated a standing order with the Cooper-Hewitt Company, in New York City, for standardized fused quartz tubes—empty or "blank" quartz bulbs that would be sent to Kunz's lab. Kunz would then "fill" or complete the tubes, which involved creating a vacuum in the tube, then distilling photosensitive metal, usually potassium, onto the inner surface of the bulb to form the cathode (electron-emitting electrode). Then the tube was filled with hydrogen and a high voltage was applied to form a thin layer of potassium hydride—the actual photosensitive element—inside the bulb. The entire process involved careful attention and a practiced eye, for example to know when the glowing surface of the cathode looked just right without going too far and ruining the sensitive surface. After the tube was cooled, a vacuum was created again and a filler gas was added, usually argon, which produced a modest amplification of the photocurrent. The tube was then sealed and ready for calibration. Kunz sent his best tubes to Washburn Observatory. Stebbins paid him twenty-five dollars for Pyrex tubes and fifty dollars for those made of fused quartz.

Jakob Kunz, a Swiss immigrant and a physicist at the University of Illinois, invented photoelectric detectors known as "Kunz tubes," which were the most sensitive light detectors available to astronomers until the advent of the photomultiplier tube. Kunz was deeply familiar with the then-novel theories of quantum physics and relativity, and he helped introduce and teach them in the United States. DEPARTMENT OF ASTRONOMY, UNIVERSITY OF WISCONSIN–MADISON

Throughout their long friendship, Stebbins relentlessly championed Kunz's career, finding funding and promoting his lab work. Stebbins had a self-interest in Kunz's lab work, of course, but he also tried to remedy what he saw as the University of Illinois's poor treatment of Kunz, to the point of trying to find him better positions at other schools. Stebbins's research helped Kunz develop the detectors in his laboratory, because it provided ever more stringent tests (it was always possible to find a dimmer star) of the various tube designs that Kunz tried. The collaboration between Stebbins and Kunz was an essential element in the development of modern astrophysics made possible by Wisconsin astronomers.

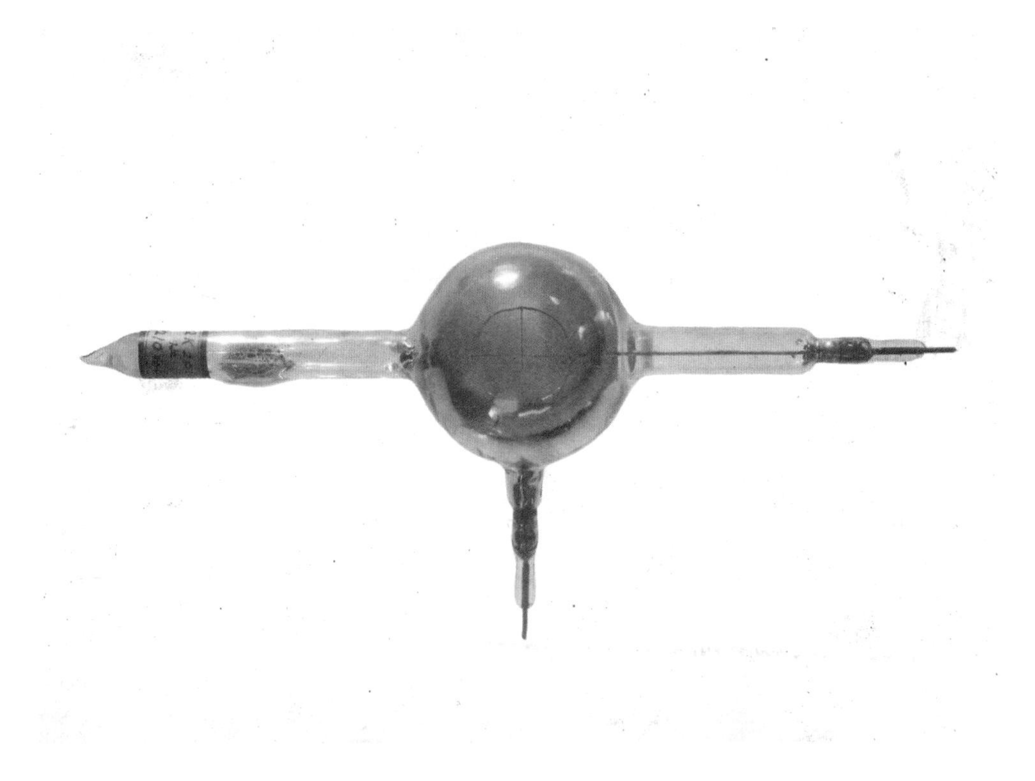

This Kunz tube, numbered QK201, was Jakob Kunz's 201st tube and was probably made in the mid-1920s. Joel Stebbins published this photo in 1927. The Q means the glass portion of the tube is made of fused quartz, which was produced by the Cooper-Hewitt Company in New York City. The K means the photosensitive coating is potassium (chemical symbol K) hydride, which Kunz distilled onto the inside surface of the bulb in his lab. The photosensitive coating serves as the cathode (or the electron emitting element) and extends through to the stem pointing down from the bottom side of the bulb. The anode (or the electron collecting element) can be seen on the face of the bulb as a ring-and-cross structure of wire, which extends through the right arm of the tube. With a few hundred volts applied between cathode and anode, a very small electrical current will flow when light (from a star, for example) reaches the potassium surface, and the electrical current will be a measure of the amount of light. The roughly spherical bulb of QK201, which no longer exists, would have measured about 50 millimeters, or about two inches, in diameter. DEPARTMENT OF ASTRONOMY, UNIVERSITY OF WISCONSIN–MADISON

This image shows one of the Stebbins-Kunz photoelectric photometers with the sides removed to reveal the internal construction, including the Kunz tube (probably QK201). A mirror in the optical path from the telescope above allowed the astronomer to center the target in the eyepiece. Rotating the mirror away allowed the light to reach the Kunz tube below. Batteries supplied the high voltage across the tube, which produced the small electrical current when starlight reached the photosensitive surface. Just visible is the fine wire connecting the tube's cathode connector to an electrometer, a very sensitive electrical meter that hung below the photometer box on a gimbal, or a kind of universal joint, to keep the electrometer vertical as the telescope moved. DEPARTMENT OF ASTRONOMY, UNIVERSITY OF WISCONSIN-MADISON

20

THE STEADFAST RIGHT-HAND MAN

CHARLES MORSE HUFFER

Charles Morse Huffer (1894–1981) was Washburn Observatory's second astronomer during the entire directorship of Joel Stebbins, just as Albert Flint had served during most of George Comstock's time. As the highly sociable Stebbins traveled to summer observing runs in California and other professional engagements around the country, Huffer (who was widely known by other astronomers simply as "Morse") kept the observatory operating. He supervised the staff, taught classes, and spent countless nights on photometry of eclipsing binary stars and countless days calculating their orbits and masses. He also managed and carried out most of the observing for the program that mapped space reddening. Huffer distilled his decades of teaching experience into a line of very successful college-level astronomy textbooks, one of his more visible achievements, but he also turned his training in the mathematics of classical astronomy to the development of some of the first orbital computation methods implemented on the earliest electronic computers. Huffer thus ensured Washburn Observatory's stability, continuity, and scientific productivity through the middle of the twentieth century until he retired from the University of Wisconsin in 1961 to pursue a second teaching career at San Diego State University.

Huffer first met Stebbins and became interested in astronomy after graduating from Albion College in Michigan and heading to the University of Illinois for graduate studies in mathematics. After finishing his master's degree at Illinois in 1917, Huffer was hired to work at Lick Observatory's Southern Hemisphere observing station near Santiago, Chile, where he remained until early 1922. As Stebbins made plans to establish his research operation at Washburn Observatory, he wrote to Huffer in January 1922, offering the job of assistant astronomer. Huffer immediately accepted and by August had relocated to Madison with his wife, Ruth Tribby, whom he had met in Santiago, and their daughter, Helen.

Morse Huffer poses in Washburn Observatory's dome in 1931, with an early photoelectric photometer mounted on the end of the telescope. Huffer and his assistants used this instrument to measure the colors of more than a thousand type B stars to map interstellar extinction in the Milky Way.
WHI IMAGE ID 18922

Early in his tenure as Stebbins's assistant, Huffer experienced personal tragedies when Ruth died suddenly in 1923, followed by the death in 1925 of his mother, who had moved to Madison to help raise Helen. Despite this, he managed to complete the first doctoral degree in astronomy awarded by the University of Wisconsin and joined the faculty as assistant professor of astronomy. In 1928, Huffer married Elizabeth Hart of Waupaca, Wisconsin, who was a mathematician and occasional astronomical computer at Washburn Observatory in her own right.

Stebbins and Huffer developed the methods and standards of the early days of photoelectric photometry and tutored astronomers at other observatories in the new technology. In 1931, for example, Huffer spent seven weeks at Yerkes Observatory, where he worked with staff on photometry using an instrument Stebbins had built for them; while Yerkes astronomer Christian Elvey came to Madison during that time to learn from Stebbins. And in the first half of 1934, Huffer and research associate Albert Whitford worked in residence at Mount Wilson Observatory as ambassadors of Whitford's newly invented amplified photoelectric photometer. Under the leadership of Huffer, Stebbins, and Whitford, Washburn Observatory astronomers were the principal authorities, developers, and practitioners of photoelectric photometry until after World War II, when new technologies led to a rapid expansion of the field.

Huffer's decades of photometric observations of eclipsing binary stars kept Washburn's 15.6-inch refractor busy and productive as a research instrument long after most similar nineteenth-century telescopes had been retired as educational tools or even scrapped altogether. But Huffer's most impressive labors came in the 1930s as Stebbins's research program on interstellar reddening and interstellar matter took shape. The effects of interstellar matter, measured by the reddening of starlight using Wisconsin's photoelectric photometers, could only be detected in distant objects like globular clusters by using the large Mount Wilson telescopes, which Stebbins frequently used during the summers. But to understand the broader distribution of interstellar matter required tens of thousands of photometric measurements of the colors of more than a thousand stars that were detectable in Madison with the Washburn refractor. Huffer made this work happen by employing and supervising teams of student observing assistants beginning in about 1932. In 1934, the *Publications of the Washburn Observatory* published the photometrically measured colors of 733 stars. By 1940 they had measured a total of more than 1,300 such stars, which made possible the first maps of the distribution of interstellar reddening in our galaxy and a better understanding of its size and structure.

In his personal time, Huffer was an accomplished musician and served as his church organist and choir director for many years. He also was an active supporter of amateur astronomers, and in the early 1930s he helped found the Madison Astronomical Society, whose meetings and activities he hosted for many years at Washburn Observatory. He also served as the club secretary until 1960, when he was elected its president. From 1946 until 1955, Huffer also served as secretary of the American Astronomical Society. Huffer would continue as second astronomer during Whitford's tenure as director, and throughout his decades-long career would remain Wisconsin's leading expert on eclipsing binary stars.

21

A NEW GENERATION IN ASTRONOMICAL PHOTOMETRY

ALBERT WHITFORD

Albert E. Whitford (1905–2002), like his mentor, Washburn director Joel Stebbins, was a major figure in the astronomical world and, as a native of Milton, Wisconsin, was one of three prominent twentieth-century astronomers from southern Wisconsin, along with Gerald Kron (1913–2012) from Milwaukee and Olin Eggen (1919–1998) from Orfordville. All three were shaped by Washburn Observatory.

Whitford graduated from Milton College, where his father taught mathematics and physics, and came to Madison to pursue a PhD in physics in the fall of 1926. In the fall of 1931, nearing the completion of his doctoral research and in search of a job, Whitford became the latest in a series of physics graduate students recruited by Stebbins to join Washburn Observatory's staff to work on instrument development.

As he had done at Urbana, Stebbins made a point of working with researchers in the university's physics department to develop new ways of solving old problems. As early as 1922, when Stebbins first arrived in Madison, he had hoped that the electronics expertise of physics professor Earle Terry, who pioneered early radio transmissions from the university, might lead to an electronically amplified photoelectric photometer. But success eluded them until Stebbins hired Whitford and assigned him the job of designing an amplifier that was sensitive to the minuscule photocurrents produced by the Kunz tube. A successful amplifier could improve the measurements of dimmer stars. The idea of using electronic amplification to improve the sensitivity of a photoelectric photometer was not particularly new, but other researchers had tried and failed to make a functional photoelectric amplifier.

Whitford succeeded by early 1932. Building on the work of his predecessors (previous physics researchers hired by Stebbins), Whitford enclosed his circuitry in a vacuum tank and pumped out most of the air. The result was an amplifier with four times the sensitivity

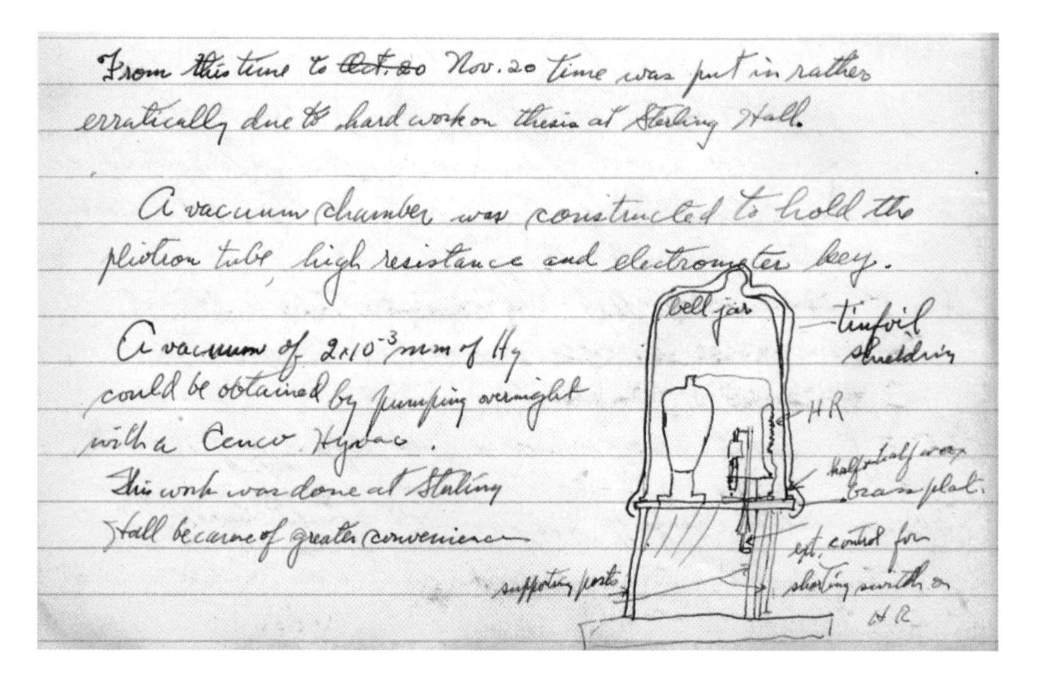
From this time to ~~Oct. 20~~ Nov. 20 time was put in rather erratically due to hard work on thesis at Sterling Hall.

A vacuum chamber was constructed to hold the pliotron tube, high resistance and electrometer key.

A vacuum of 2×10^{-3} mm of Hg could be obtained by pumping overnight with a Cenco Hyvac.

This work was done at Sterling Hall because of greater convenience

Albert Whitford recorded a pivotal development in Wisconsin astronomy when he made this entry in his lab notebook in the fall of 1931, describing his invention of the amplified photoelectric photometer. Working in a laboratory in Sterling Hall, Whitford enclosed a DC amplifier tube (also called a "pliotron," itself a recent invention by the Radio Corporation of America) with other photometer circuitry under a bell jar. When the air was pumped out of the bell jar with a vacuum pump, the erratic electrical "noise," which had plagued previous amplifiers, was reduced to a tenth of its previous level. Whitford immediately began constructing a practical version that could actually operate on the Washburn Observatory telescope, which he put into operation in April 1932, dramatically increasing the power of the old telescope to detect dim stars. UNIVERSITY OF WISCONSIN–MADISON ARCHIVES, MADISON, WISCONSIN

of Stebbins's electrometer instrument. The new instrument was so effective, Stebbins liked to say, that it was as if they had doubled the size of the Washburn telescope at a small fraction of the cost. The amplified photoelectric photometer immediately opened up new frontiers of astronomical investigation.

Whitford completed his PhD in physics in the summer of 1932, but he found few job prospects because of the Great Depression, so he stayed at Washburn and continued working for Stebbins as an observatory assistant dedicated to improving the new instrument. Stebbins found support for Whitford's position through grants from the National Research Council and the Carnegie Foundation via George Ellery Hale, who wanted photometry developed for his Mount Wilson Observatory and for the new 200-inch telescope then in development. This indicates how highly the broader scientific community valued the work at Washburn

Observatory. By the summer of 1933, Stebbins had arranged a one-year fellowship for Whitford at Mount Wilson, where he carried out research under Stebbins's direction, using his amplified photoelectric photometer on the California observatory's large telescopes. On the mountaintop, where photographic plates were the more familiar technology, Whitford became known for keeping a soldering iron at the ready for working on his electronics, a field unfamiliar to the traditionally trained astronomers. Under Stebbins's tutelage, the physicist became an astronomer; Whitford was perhaps the first scientist to singularly unify the worlds of modern physics and modern astronomy.

By February 1933, Albert Whitford had built the second of his amplified photometers specifically for use on the 100-inch telescope at the Mount Wilson Observatory, then the largest telescope in the world. Stebbins and Whitford took the instrument to Mount Wilson for the summer observing season, when this photo was taken. The image shows Whitford at his operating position at the foot of the northern pier of the giant telescope. The brass photometer vacuum tank was far overhead, mounted at the Newtonian focus of the telescope. The photometer's controls and galvanometer scale, used to read out the photometric measurements, occupy the small wooden table. Thus began a new epoch in astronomy in which observers could operate their instruments at a distance from their giant telescopes. DEPARTMENT OF ASTRONOMY, UNIVERSITY OF WISCONSIN–MADISON

Stebbins kept Whitford employed as a research associate at Washburn Observatory for several years, which effectively served as Whitford's astronomical apprenticeship. For an apprentice, the scientific rewards were enormous, as Whitford was able to participate in some of the most far-reaching astronomical research ever to come out of Wisconsin. In those years, Whitford worked closely with then student assistant Gerald Kron as the two developed new applications for photoelectric photometry. They even obtained prototype photomultiplier tubes from the Radio Corporation of America more than a decade before those tubes would become widely available. And when they found those early detectors too noisy for scientific measurements, Whitford and Kron instead used the new technology to make the world's first automatic electronic telescope guider.

Whitford became an assistant professor and Washburn staff astronomer in the fall of 1938, after Stebbins convinced university president Clarence Addison Dykstra to expand the observatory staff from two astronomers, as it had been since 1889, to three. There would not be another staff expansion, to add a fourth astronomer, until late 1958.

Meanwhile, the war in Europe was spurring the United States to mobilize its scientific talent, and by the end of 1940, Whitford was recruited by the MIT Radiation Laboratory to conduct military research. He and his wife, Eleanor Whitelaw, whom he had married in 1937, moved to Massachusetts, where Whitford spent the war years carrying out secret radar-related research while on leave from the University of Wisconsin. Whitford did get away from war work from time to time, for example to assist Stebbins with observing at Mount Wilson and to attend a National Academy of Sciences meeting in Madison. By the summer of 1946, Whitford had returned to his position at Washburn Observatory despite offers from other universities and private firms. Stebbins, now nearing retirement, groomed Whitford as his successor and exercised his influence to ensure that Whitford would succeed him as observatory director in the summer of 1948.

Whitford would serve as Washburn Observatory's director for ten years, continuing and extending the research and instrument development activities that he and Stebbins had begun. In 1954, Whitford was elected to the National Academy of Sciences—just like Stebbins, Comstock, Holden, and Watson before him. Aside from his research work, much of Whitford's directorship would be spent navigating the postwar challenges that would transform Washburn Observatory. It fell to Whitford to manage the creation of a department of astronomy, initiated by Stebbins but not approved until after his retirement, within the College of Letters and Science. The university's postwar enrollment explosion and consequent growth in faculty considerably amplified what might otherwise have been a mostly bureaucratic reorganization. The relocation of the new department from the old observatory to its new quarters in Sterling Hall was another part of the great postwar changes. Whitford also managed the financing, planning, and construction of the new Pine Bluff Observatory, the "country" telescope that Stebbins had spent years promoting.

Whitford's last day with the university was the day of the dedication of Pine Bluff Observatory, June 30, 1958, but he was not retiring. He had been recruited by the University of California to be director of Lick Observatory, where Stebbins was now a senior research fellow and former Washburn student Gerald Kron was a senior staff astronomer. Lick Observatory, where Edward Holden had pioneered the modern astronomical reflecting telescope, was building a huge new 120-inch reflector, and they needed Whitford, another Wisconsin astronomer who knew how to build an observatory, to finish the project.

Albert Whitford (left), director of Washburn Observatory, and Morse Huffer examine the mirror blank that would soon be shipped to the Yerkes Observatory optical shop, where it would be worked into the primary mirror for the Pine Bluff Observatory 36-inch telescope. DEPARTMENT OF ASTRONOMY, UNIVERSITY OF WISCONSIN–MADISON

22

FOUR GENERATIONS OF PHOTOMETERS

In the first half of the twentieth century, Washburn Observatory astronomers earned fame for employing and advancing the technique of photoelectric photometry, one of the foundation technologies of modern astrophysics. Photometry is the measurement of light. An astronomer needs photometry to answer the basic question, "How much brighter is this star than that one?" One way of measuring the brightness of stars is to take advantage of something called the photoelectric effect. The photoelectric effect allows an observer to measure light intensity by measuring the number of electrons that are emitted from a special metal surface when light is absorbed by it.

Photoelectric devices were key to Wisconsin's prominence. Joel Stebbins arrived in 1922 with nearly a decade of work experimenting with and developing photoelectric photometry at Illinois, which he immediately put to use at Washburn Observatory. Stebbins built an entirely new and somewhat improved astronomical photometer, based closely on the ones he had built at Urbana, which was operational by the summer of 1923. The light from the telescope passed into a box that contained the light-sensitive photoelectric cell. The very weak electric current the photoelectric cell produced from starlight could only be measured by a sensitive device called an electrometer, which hung vertically from the photocell box. This type of electrometer, called a string electrometer, had to remain vertical at all times regardless of the motion of the telescope, so it was suspended from a kind of universal joint and weighted to keep it plumb. To observe, the astronomer peered into the eyepiece to see not the star, but the fine "string" (usually a very fine platinum wire or a gilded quartz fiber) whose slight deflection measured the intensity of the starlight. Though it represented an improvement over its predecessors, the string electrometer was a temperamental device, and its awkward presence limited the range of motion of the telescope. The photometer box was made to Stebbins's specifications in the university machine shop.

The other major component, the string electrometer, was a commercial product, which Stebbins had to modify extensively to bring it up to his standards.

By summer 1927, British physicist Frederick Lindemann invented a new kind of electrometer that was just as sensitive but far more compact, thus eliminating the need for the ungainly string electrometer. In addition to its quite small size, the Lindemann electrometer would work in any orientation, so it could move anywhere the telescope could point. Though still fragile, the tiny Lindemann electrometer meant the photometer could readily be taken to other telescopes, and Stebbins had it working on Mount Wilson's 60-inch reflector by August 1930 and on the 100-inch, the world's largest, by September 1931. Despite the increased versatility of the new instrument, the astronomer still examined the measurement scale by peering through the microscope to read the electrometer, now mounted on the photometer box itself, and therefore had to be on duty at the eyepiece the entire time the telescope was observing a target. This could put the observer rather high above the cold, dark, very hard floor under the big telescopes at Mount Wilson. The rigors of this kind of astronomy were no joke.

Joel Stebbins with Washburn Observatory's earliest photoelectric photometer, which became operational in about 1923, mounted on the 15.6-inch equatorial telescope. The rectangular box contained a Kunz tube, which was the actual light detector. Suspended from a gimbal joint below the box hangs the string electrometer. The observer read the scale of the electrometer by peering into the microscope, seen extending to the left from the electrometer body toward Stebbins. UNIVERSITY OF WISCONSIN-MADISON ARCHIVES, MADISON, WISCONSIN

Another major advance was in place by the summer of 1932. Stebbins had hired the young physicist Albert Whitford in late 1931 to develop a way to amplify the feeble photocell currents. Whitford solved the problem by isolating the photocell and an amplifier tube inside a vacuum tank, so the new instrument was housed in a cylindrical brass vacuum tank on the end of the telescope. The electrical signal from the amplifier was strong enough to deflect the indicator of a common laboratory electrical measuring device called a galvanometer, which could be connected by wires at some distance from the telescope, replacing the troublesome electrometers. Freed from the electrometer, the astronomer no

The photometer seen here sports the new Lindemann electrometer. Gone is the pendulous string electrometer. The microscope for reading the scale on the Lindemann electrometer was larger than the electrometer itself, but the astronomer still had to be stationed at the end of the telescope to get the measurement.
DEPARTMENT OF ASTRONOMY, UNIVERSITY OF WISCONSIN–MADISON

Stebbins sits at the controls of the new Whitford photometer, where he can read (on the horizontal screen held in the lab stand in front of him) the deflection of the galvanometer and record the numbers in his notebook. The astronomer was now freed from the eye end of the telescope. Behind Stebbins stands Edwin J. Bernet, an advanced student observing assistant and electronics technician (a new niche in astronomy) who worked at Washburn Observatory from about 1933 until 1937. DEPARTMENT OF ASTRONOMY, UNIVERSITY OF WISCONSIN–MADISON

After World War II, the venerable Washburn refractor remained on the photometric frontier as one of the first telescopes to be equipped with the photomultiplier tube (or PMT) outside of Lick Observatory, where Stebbins's former student Gerald Kron built the first one. Here, Morse Huffer operates the early PMT photometer. The signal from the PMT was sufficient to drive a chart recorder, just visible centered in the background at the bottom. The recorder provided a direct record of the photometer output in place of galvanometer readings handwritten in a notebook, as was common practice with Whitford's instrument.
DEPARTMENT OF ASTRONOMY, UNIVERSITY OF WISCONSIN–MADISON

longer had to be present at the focus of the telescope to make a photometric reading and could instead sit at a table on the dome floor to record the galvanometer readings. Far more importantly, with Whitford's new amplified photometer, it was possible to measure stars nearly six times dimmer than before. For the cost of Whitford's salary, Wisconsin astronomers had made the 100-inch telescope perform as if it were a 200-inch telescope, for photometry at least.

The Whitford photometer remained the most sensitive astronomical photometer in use until after World War II. In the mid-1930s, the Radio Corporation of America (RCA) invented a new light-detecting device called the photomultiplier tube, or PMT. It would be nearly impossible for most astronomers to obtain the new device until the postwar period. Stebbins, however, carefully cultivated strong relations with RCA, so Washburn astronomers, even before the war, became the first astronomers to experiment with prototype PMTs. The PMT would prove to be far more sensitive than the Whitford instrument, was even more compact, and didn't require a vacuum pump.

By the late 1940s, the Washburn 15.6-in telescope was one of the first in the world to sport a working PMT photometer. By the end of his career, Washburn astronomer Morse Huffer had measured eclipsing variable stars on the Washburn telescope with every generation of the Washburn photoelectric photometers from 1923 until 1958.

23

MECHANICAL UPGRADES

REMOUNTING WASHBURN'S TELESCOPES

Both of Washburn Observatory's equatorial research refractors, the 15.6-inch and the six-inch, were made by Alvan Clark and Sons. Clark also made the equatorial mountings, which are a very important part of an astronomical telescope. The two mountings bore a strong resemblance to one another, aside from the much heavier construction used for the larger telescope. Equatorial mounts were the standard design for professional astronomical telescopes until roughly the 1990s, when computerized altitude-azimuth mounts for large telescopes became typical. Equatorial in this case means that the mount can move the telescope in a plane parallel to Earth's equator, so that the telescope can easily track stars across the sky. To do that, the polar axis of the mount is built at a fixed angle, relative to the local horizon and determined by the latitude of the observatory, so as to point to the north celestial pole, around which the stars appear to rotate. Washburn's six-inch refractor was originally built for Sherburne Burnham when he lived in Chicago, so when Burnham relocated to Madison, the base of the mount had to be shimmed up slightly to adjust to Washburn's more northerly location.

The designer of an equatorial mount must figure out how to transmit torque to the polar axis, smoothly and at just the right speed. Clark solved this problem by using a sector arm clamped at its upper end to the polar axis. At its lower end the arm was connected to a set of gears operated by a clockwork drive. The arm had a limited run and could only drive the telescope for roughly an hour or so before the astronomer had to unclamp and reset the arm, which created an unwelcome and tedious interruption to the astronomer's work.

Another inconvenience was that telescopes could be pointed only by pushing or pulling from one end or the other. But with a long refractor like the 15.6-inch, the ends of the tube often could be well out of human reach, so ropes were kept tied to its extremities in order to tug it this way or that. Repointing the telescope in a significantly different direction—

The new mounting for the Burnham six-inch telescope shows Oscar Romare's innovations, including the hour angle indicator and polar axis wheel. (To better appreciate the difference, compare it with the original Clark mounting pictured in chapter 5 in the photograph of Sherburne Burnham's six-inch refractor.) The telescope mounted here in the Student Observatory is the ten-inch refractor, whose objective lens was made by J.W. Fecker. Stebbins mounted it in place of the Burnham in 1927 for photometry experiments.
DEPARTMENT OF ASTRONOMY, UNIVERSITY OF WISCONSIN–MADISON

from the east side of the dome to the west, for example—required dexterous maneuvers to handle the ropes safely. It was a complicated operation, somewhat like managing a sailboat.

Over time, it comes as no surprise that many astronomers grew weary of these troublesome routines, which could be particularly disruptive in the long sequences of measurements that became typical in Joel Stebbins's photometric work. Stebbins wanted to remount the 15.6-inch telescope as soon as he arrived in Madison, but it was a costly proposition. For example, Warner and Swasey, who had made mountings for the Lick and Yerkes Observatories, would charge about $15,000 for a new equatorial mounting. That was a great deal of money in 1922 and not an outlay the university could afford.

As luck would have it, the university's machine shop had hired Oscar E. Romare (1875–1932) in 1920. Romare was born in Sweden and spent about eighteen years as a machinist and instrument maker at Yerkes Observatory before coming to Madison. He had a deep knowledge of astronomical engineering and took an immediate interest in Washburn Observatory, just as Stebbins was getting started. To begin with, Romare built, to Stebbins's specifications, the first in a long line of photoelectric photometers for the observatory, and soon he built another for Yerkes, also designed by Stebbins.

The cast-iron base of the pier for the new Romare mounting is being hoisted up so it can be passed through the slit in the observatory dome for installation. On the ground, leaning against the building, rests the original Clark mounting. DEPARTMENT OF ASTRONOMY, UNIVERSITY OF WISCONSIN–MADISON

Stebbins invited Romare to create a replacement for the old telescope mountings. Romare designed one that was strongly influenced by the Warner and Swasey mounting he knew from Yerkes. The new mounting was supported by a hollow cast-iron pier that could accommodate the clock drive within, unlike the masonry piers used to support the Clark mountings. Romare's design replaced the Clark sector arm drive with a gear arrangement that could drive the polar axis continuously, limited only by the telescope's range of motion. He also provided a wheel, located on the north side of the pier, to allow the operator to move the

Workers prepare to hoist the polar axis frame and bearing assembly of the new Romare mounting up and into the dome for installation on the new pier. DEPARTMENT OF ASTRONOMY, UNIVERSITY OF WISCONSIN–MADISON

telescope eastward or westward from horizon to horizon while standing in one spot, thus eliminating the need for ropes.

Another Romare innovation was a dial above the wheel, allowing the operator to directly read the angle of the telescope east or west of the meridian (which astronomers call the hour angle) as the wheel and driving clock rotate the telescope. The new design also included electrical controls for the clutch and slow motion of the polar axis drive. Romare's design for the telescope's driving clock itself was also innovative in its use of ball bearings, and the entire design was considered highly advanced for its day.

These all would be major improvements over the Clark mounting, and Romare and machinist M. H. Kidder built them in the university shop at about one-third the cost estimated by Warner and Swasey—an amount that Stebbins was able to afford thanks to a few small grants of university funding. Before modifying the 15.6-inch telescope, Stebbins tested Romare's design by having him make a smaller version to replace the Clark mount for the Burnham six-inch telescope in the Student Observatory. The new mounting for the Burnham refractor was installed by May 1927, after which Stebbins approved work to begin on the bigger telescope. Burnham's original Clark mounting, as well as its original

The 15.6-inch Washburn refractor on the new Romare mounting. On the north (right) side of the pier are the polar axis motion wheel, which looks like a ship's helm, and, above it, the hour angle circle. The lower end of the telescope is equipped with an advanced version of Whitford's amplified photometer. DEPARTMENT OF ASTRONOMY, UNIVERSITY OF WISCONSIN–MADISON

wooden tube, went back to Chicago in early 1930 on loan to the new Adler Planetarium as part of its opening exhibition.

Romare's skills as an astronomical designer were in demand beyond Madison, which unfortunately led to his tragic end. Romare had designed a small telescope installation for St. Norbert College, in De Pere, Wisconsin, just as the new mounting for Washburn's 15.6-inch telescope was nearing completion. On April 3, 1932, as he was driving to De Pere to inspect the work, Romare was killed when his car slid off an icy road. Despite this grim setback, Kidder finished the work Romare had begun and supervised the installation of the new telescope mount between April and June 1933. By July, the old Clark telescope, now more reliable and easier to operate, was back in harness and observing again with the new Whitford amplified photoelectric photometer.

24

EXPLORING STAR SYSTEMS OF ECLIPSING BINARIES

Our Sun orbits through the disk of our galaxy accompanied by Earth and the rest of the solar system, but without any companion stars. This puts it in the minority, as most stars are found in systems of several stars. Our Sun is also fairly constant in its luminosity, or the amount of energy it radiates into space, unlike many stars in the sky. Many are variable stars, so named because their brightness can vary (or appear to vary), sometimes at regular intervals, sometimes irregularly, on time scales from hours to years.

Two or more stars orbiting each other, which astronomers call a binary star system, can be detected in many ways, including directly seeing their motions through a telescope. In some cases, if the astronomer's view of their orbit is close enough to edge-on—that is, if the observer's line of sight passes close enough to the star system's orbital plane—one star will periodically cross in front of the other, causing eclipses. In the most dramatic cases, such as the variable star Algol in the constellation Perseus, the sudden dip in brightness this creates can be seen with the naked eye, but the effect is usually much more subtle.

Detecting and making sense of the slight variations in starlight caused by such eclipses was Joel Stebbins's first major scientific success. By bringing his techniques of astronomical photometry and his photoelectric detector technology to Washburn Observatory, Stebbins opened a new chapter in Wisconsin's astronomical history and in the study of eclipsing binary stars.

To discover such star systems, astronomers make a series of brightness measurements of a star suspected of being an eclipsing variable. In Stebbins's era, such measurements almost always had to be made relative to one or more comparison stars, which meant a great many observations were necessary to construct a brightness curve showing how the star's brightness varied. Eclipses were the alternately deep and shallow "valleys" in the light curve. While Stebbins was the pioneer of the photometric techniques used on variable

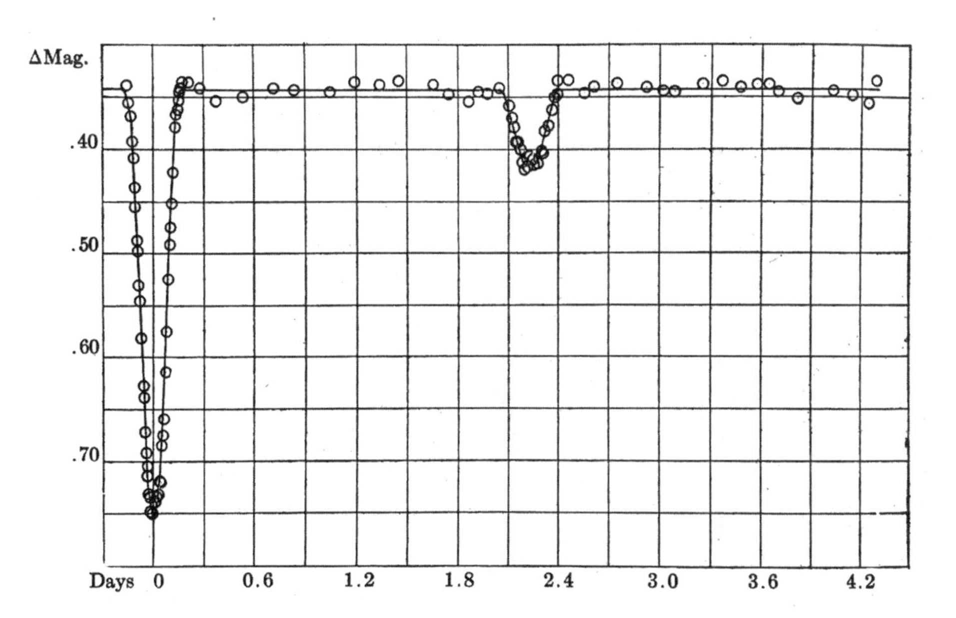

This graph showing the "light curve" of the eclipsing variable star 21 Cassiopeiae (also called YZ Cassiopeiae) was published by Joel Stebbins and Morse Huffer in 1928. 21 Cassiopeiae, which looks like a single star even in a telescope, is really two stars gravitationally bound, called a binary star. The changing brightness of the star (really the combined light of both stars) is measured in units of astronomical magnitudes and plotted on the vertical axis; while time, measured in days, is plotted on the horizontal axis. Each small circle represents a measurement of the star's brightness, which is an average of repeated measurements relative to nearby comparison stars, which are assumed to be constant in brightness.

Stebbins measured the star on forty nights, while Huffer measured it on forty-five nights. Each circle required eight to ten measurements of target and comparison stars, which means this brightness curve represents hundreds of individual measurements at the telescope using the Kunz tube photometer and string electrometer setup, followed by many hours of computation done by hand (by staff computers and the astronomers themselves). The deep plunge in brightness at Day 0 represents the primary eclipse, when the dimmer star eclipses the brighter of the pair. A little more than two days later, the brighter star eclipses the dimmer star, producing the secondary eclipse. Analyzing the relative depths and widths of the eclipses reveals important characteristics of the individual stars and the star system. *PUBLICATIONS OF THE WASHBURN OBSERVATORY* (1928)

stars, his student and colleague Morse Huffer was Wisconsin's most prolific eclipsing binary observer and expert from the 1920s into the 1950s.

From the brightness curve and the distance to the star system (found by stellar parallax, Albert Flint's specialty), astronomers deduce important information about the star system, the individual stars, and even the internal structure and possible interactions of the stars themselves. Their orbital motions reveal the stars' masses, and the eclipses reveal their relative sizes. The size and mass give the density of a star, and the shapes of the eclipses reveal the surface brightness variation across the disk of the stars (called limb darkening in astronomical jargon). Stebbins and Huffer both devoted many years to photometrically observing the brightness curves of eclipsing systems, analyzing the data to understand star systems and derive the physical characteristics of stars. Their work helped lay the foundations of modern stellar astrophysics and led to a wide range of applications, from the discovery of exoplanets to the study of binary asteroids.

Stebbins relentlessly pushed the improvement of photometric technology to make Washburn's instruments more sensitive and versatile. The photometry of eclipsing binaries kept the 15.6-inch Clark refractor in active use producing scientific results well past the time when most telescopes of that vintage had been retired from research. In fact, despite being located in the middle of a growing campus and city, Huffer was measuring brightness curves of eclipsing variables with the 15.6-inch telescope well into the 1950s. By then, he could analyze the data and solve for the orbits of his binary star systems using an early electronic computer, which, though primitive by today's standards, never could have been dreamed of in 1878 when the Washburn telescope was new.

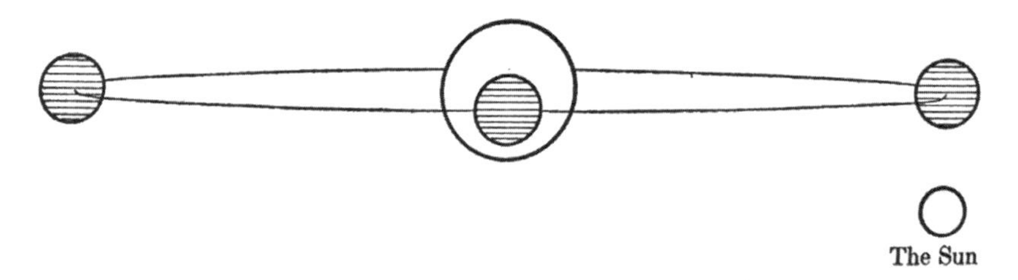

Based on their photometric light curve, Stebbins and Huffer reconstructed the sizes and masses of the eclipsing binary stars of 21 Cassiopeiae using an orbit derived spectroscopically by Canadian astronomer J. S. Plaskett and using work by A. S. Eddington relating a star's spectral type (a way of classifying stars based on their temperature and the light they emit) to its mass. Stebbins and Huffer concluded that the mutually eclipsing stars are 3.4 and 1.6 times as massive as our Sun, and 3.1 and 1.5 times our Sun's diameter. By comparison, modern figures, which are considerably more precise, show the stars to be 2.308 and 1.325 times as massive as our Sun, and 2.547 and 1.359 times our Sun's diameter. (Most of the discrepancy comes from an improved understanding of the relationship between stellar masses and spectral types.) The study of star systems helps us understand the physical characteristics of stars and is fundamental to our modern understanding of star formation and evolution. *PUBLICATIONS OF THE WASHBURN OBSERVATORY* (1928)

25

ECLIPSE EXPEDITIONS

SOLAR CORONA PHOTOMETRY

Because of the Moon and the eclipses it causes, our Sun is, in effect, an eclipsing variable star—at least as seen from Earth. So it seems fitting that under Joel Stebbins, total solar eclipses became yet another area of expertise for Wisconsin astronomers, because photoelectric photometry could provide quantitative measures of the brightness of the Sun's corona—the outermost part of the Sun's structure, which is only visible during a total solar eclipse.

Stebbins, who had a talent for asking good research questions, saw one that his earliest photoelectric photometer might answer: does the light from the Sun's corona vary along with the well-known sunspot cycle—and if so, how? Sunspots, or areas where the Sun's magnetic field is particularly strong, come and go on the Sun's surface. Since the seventeenth century, scientists have known that the number of sunspots rises and falls in cycles, roughly every eleven years. The brightness of the solar corona was suspected of varying in a similar cycle, but quantifying such changes was very tricky, so Stebbins planned to see what his photometric techniques could reveal.

To investigate that question, Stebbins and his longtime collaborator, physicist Jakob Kunz, organized their first eclipse expedition while still colleagues at the University of Illinois. They took photoelectric photometers to the area outside of Rock Springs, Wyoming, into the path of totality for the solar eclipse of June 8, 1918. Using their photometers, Stebbins and Kunz were able to measure the brightness of the solar corona and found it was equal to roughly ten percent of the brightness of an equivalent area of daylight sky. Most importantly, they proved their equipment was capable of the measurements they had set out to make.

Their first expedition was so successful that Stebbins, after moving to Washburn Observatory, organized expeditions for the total eclipses of September 10, 1923, at Catalina Island, off the California coast, where they were clouded out; January 24, 1925, at Middletown, Connecticut, where it was clear; August 31, 1932, at Lancaster, New Hampshire, where

Joel Stebbins designed the equipment shown here for photometry of the solar corona during total solar eclipses. This setup likely was used for the solar eclipses of June 1918, September 1923, January 1925, and August 1932. (The equipment Stebbins used in Peru for the June 1937 eclipse was far simpler and more compact.)

On the upper left are two tripod-mounted open tubes (not telescopes, despite the resemblance), each of which were attached to a photometer box containing a Kunz tube. To the left of the tripod base is the box containing a "standard candle" light source (for calibration), which would cast light through the wide tube connecting it to a photometer box, partially concealed here. On the right, a small cabinet holds the two cylindrical galvanometers, one for each photometer, that register the light measurements. The galvanometers worked thanks to a small mirror that could swing on a fiber and deflect a beam of light. The galvanometers required a dark location inside a shelter but near the photometers, which had to remain outdoors for a clear view of the Sun. On the table on the lower right are essential accessories such as stopwatches, a metronome to assist in timing, a corked flask of amyl acetate used as fuel for the standard candle, and the ever-present notebook for recording data. DEPARTMENT OF ASTRONOMY, UNIVERSITY OF WISCONSIN–MADISON

they were clouded out again; and, finally, June 8, 1937, at Cerro de Pasco, Peru, where it was clear. Throughout these expeditions, Stebbins and Kunz continued using photoelectric photometers to measure the brightness of the Sun's corona. They found that the coronal brightness was effectively equal during the 1918 eclipse, around the time sunspots would have been at their maximum, and the 1925 eclipse, when sunspots would have been at their minimum. Their measurements during the 1937 eclipse again showed no significant variation in coronal brightness. In addition, Stebbins and Kunz measured the color of the

Jakob Kunz (near center), Joel Stebbins, and other Washburn Observatory colleagues worked from this encampment in Lancaster, New Hampshire, during the solar eclipse of August 31, 1932. The small shingled building contained the galvanometers and other equipment, except for the twin tubes and their attached detector boxes, which can be seen at far left. During totality, Kunz would have been inside the building to operate the photometers and record their readings, while Stebbins would have been outside to keep the tubes aimed at the solar corona. Though everyone was looking hopefully skyward in this photo, taken about thirty minutes before totality, the eclipse was, in fact, clouded out—demonstrated by the general lack of distinct shadows in the photo. DEPARTMENT OF ASTRONOMY, UNIVERSITY OF WISCONSIN–MADISON

corona's light and ruled out the theory that coronal light was simply sunlight scattered from fine dust surrounding the Sun.

Their eclipse photometry equipment was unique. No lenses or other real optics were needed. Each Kunz tube, of which there were two for redundancy, acted as a light detector and received light through a simple long tube that isolated the spot of sky where the corona would be during the eclipse. These long sighting tubes look deceptively like telescopes because they were mounted on a tripod so they could be pointed. The spot of sky selected by the tubes was large enough to encompass the motion of the Sun during the few minutes of totality, so no clock drive was needed to follow the target.

Each Kunz detector was mounted in a box that could be removed from the sighting tubes and placed at the end of yet another horizontal tube for calibration. The calibration tube was in several telescoping sections so that its length could be varied, and it had a standard candle (a laboratory calibrated lamp) at the far end. The light from either the standard candle or the solar corona was vastly brighter than starlight, bright enough that the electric current from the Kunz detector could be measured with a laboratory galvanometer rather than the temperamental (but much more sensitive) electrometers that had to be used for stellar photometry before Whitford's amplifier became available.

The astronomical crew from Washburn Observatory attending the 1932 eclipse consisted of Stebbins, Morse Huffer, Albert Whitford, and Whitford's friend Neill Whitelaw. (Five years later, Whitford would marry Eleanor, Neill's sister.) The clouds never parted before totality, so they gathered no data on this expedition, but Stebbins nevertheless did make one discovery: he liked the maple syrup from New Hampshire. He may have returned to Wisconsin with some packed into his yellow Model A Ford alongside the scientific equipment. The following year, he wrote to a Mrs. Lucas in Lancaster, who had rented rooms to the eclipse crew, asking to purchase one gallon of her syrup to be shipped to Kunz in Urbana, and two to be shipped to himself in Madison.

26

MAPPING THE INTERSTELLAR MEDIUM WITH SPACE REDDENING

Probably Joel Stebbins's most far-reaching astronomical research program began in about 1930, when he applied his technique for measuring the effective color of a star using photoelectric photometry to stars in selected parts of the Milky Way. This would soon transform scientific understanding of the interstellar medium—the term for the gas and dust particles of matter that fill the space between stars. Because interstellar dust reddens the light of distant stars, quantitative measurements of the reddening of thousands of stars allowed Wisconsin astronomers to map the distribution and strength of such reddening in the Milky Way. Not only was this the beginning of quantitative studies of the interstellar medium (a research strength of Wisconsin astronomers to this day), but it also allowed Stebbins to approximately calculate the true size of our Milky Way Galaxy, the extent of which had been both vastly underestimated and overestimated by previous researchers, who had failed to take the light-absorbing effects of interstellar matter into account.

In the early twentieth century, astronomers faced a major question: does light freely traverse the vast spaces between the stars, or does there exist any interstellar matter affecting the light we receive from the stars? Photography showed the existence of starless regions in the sky, but were they empty voids, gaps in the general distribution of stars, or were they opaque clouds of matter, blocking light from more distant starfields? Informed judgments of astronomers soon converged on the latter, namely that there are deep, dark clouds here and there in the heavens. But astronomers still weren't sure whether these dark nebulosities were local, isolated phenomena in a generally transparent universe, or if this interstellar matter was also spread more thinly across the stellar system, with correspondingly subtle, but widely distributed, effects.

In 1909, George Comstock speculated that the relative scarcity of stars north and south of the Milky Way indicated an optically dense and dusty interstellar medium above and below the Milky Way, through which our Sun and neighboring stars were moving. If true, then understanding the general extent of the interstellar medium would be essential to understanding the structure of the universe. But other astronomers, in contrast, concluded that aside from obvious local dark clouds, the effects of interstellar matter, if it existed, could be generally ignored. Both of those theories were eventually proved incorrect, but they show how astronomers were actively grappling with the question of whether gas and dust could have a general dampening effect on the measurable light from distant sources (a phenomenon known as interstellar extinction). The outcome of the question was crucial in understanding the distances and distribution of stars, star clusters, galaxies, and other celestial objects.

By 1930, Lick Observatory astronomer Robert Trumpler had used photography to demonstrate subtle effects of interstellar matter on the light that reaches us from open star clusters. This confirmed that the effects of the interstellar medium were widespread and that figuring out how to measure those effects was essential to advance the understanding of our galaxy. Stebbins knew Trumpler well and immediately recognized that his own specialty of photoelectric photometry could be applied to the problem. By measuring the light of a given star twice, once through a yellow filter and then again through a blue filter, Stebbins could arrive at a "color index"—in effect, a quantitative measurement of stellar color. Reddish stars would be comparatively dim through the blue filter relative to the yellow filter, and bluer stars would be comparatively bright relative to the yellow. The photoelectric photometer could measure those color differences precisely.

Stebbins hypothesized that some fraction of the interstellar matter would be fine dust, comparable to the dust in our atmosphere that reddens the light of the setting Sun. The amount of reddening, if any, should be related to the total amount of extinction. To test the idea, Trumpler sent Stebbins a list of stars, called type B stars, which were known from their spectra to be very luminous, hot, bluish stars. Stebbins and Morse Huffer used the Washburn refractor to observe Trumpler's stars and measure their colors photometrically. They found the light of many of them was measurably reddened compared to the normal color of a type B star. This was independent confirmation of Trumpler's earlier finding and also took it a step further: interstellar matter was real, some of it was dust, and its effects could be quantified and mapped by photoelectric photometry.

So began Washburn Observatory's extensive program of mapping the interstellar medium. The laborious photometry of hundreds of type B stars—and later more than a thousand—was done in part by Stebbins, but mostly by Huffer and student assistants using the Washburn 15.6-inch telescope equipped with a Kunz tube and a Lindemann electrometer. Their goal was to use the type B stars, which are widely spread through the Milky Way and

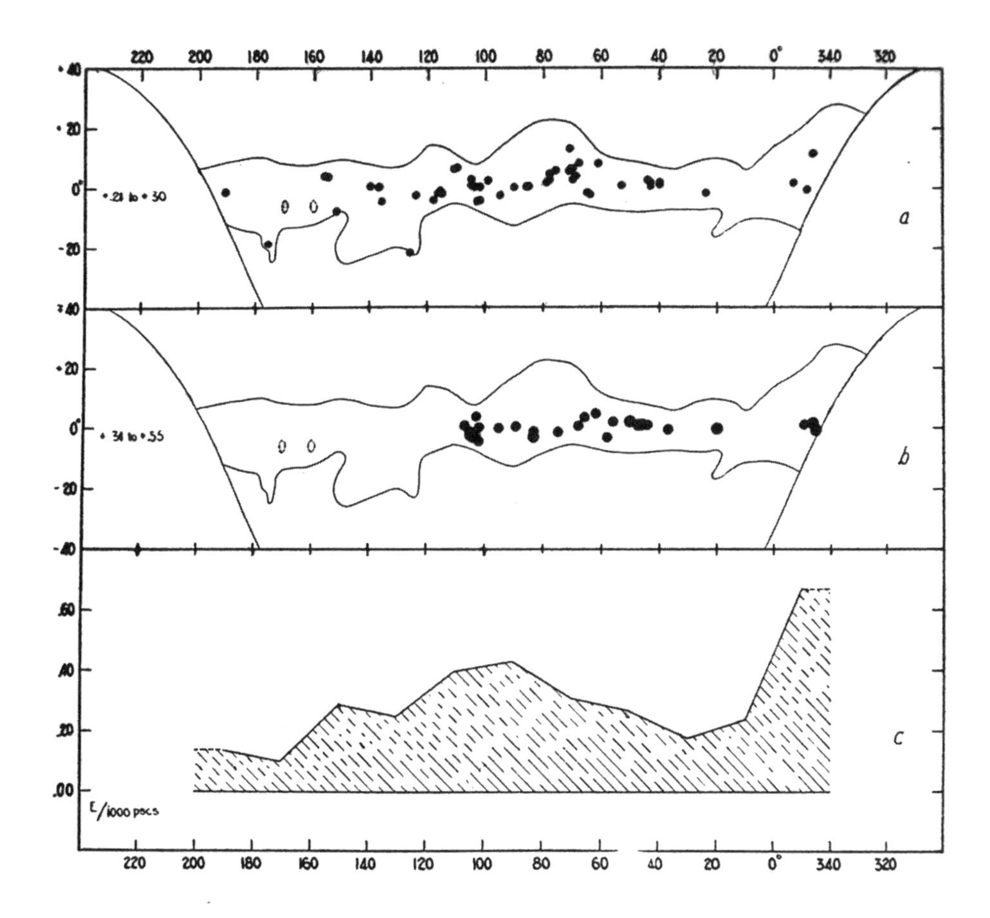

These diagrams show the early results of measuring the reddening of type B stars in the Milky Way Galaxy. The top diagram shows that reddened type B stars (represented here by black dots) are generally confined to the Milky Way, the borders of which are indicated here by the irregular horizontal lines running left to right. The zones at the left and right edges of the plot represent the southern parts of the sky that are not visible from Washburn Observatory.

The middle diagram shows that the most strongly reddened type B stars coincide closely with the galactic equator. And the bottom diagram shows that the amount of reddening varies irregularly in galactic longitude (the left-right axis of the charts) but begins to increase dramatically approaching the center of the Milky Way, just off the left edge of the plot and not visible from Madison.

Joel Stebbins published these results in the very last issue of the *Publications of the Washburn Observatory*, volume 15, which he released in 1934. The observatory's publication series had appeared frequently since Holden established it in 1881, but scientific publication patterns were changing rapidly by the early twentieth century. After Stebbins terminated the series, all subsequent scientific results from Washburn Observatory would be published in professional periodicals, such as *Astrophysical Journal*.

PUBLICATIONS OF THE WASHBURN OBSERVATORY (1934)

luminous enough to be seen across great distances, to measure the distribution and effects of the interstellar matter.

By 1934, Stebbins and Huffer were able to publish hard-won data on 733 stars, revealing for the first time the extent of space reddening, which tracks the interstellar matter as well as the varying intensity of its effects throughout the Milky Way. Based on these new results, Stebbins could show that distances to globular star clusters in our galaxy had been overestimated by factors of two or even three. The use of photometry to measure space reddening also showed that the interstellar medium was very patchy in its effects, but also nearly ubiquitous in the plane of the Milky Way—the regions of the sky outlined by the starfields of our galaxy. In addition, Stebbins showed that the extinction of starlight was mostly caused by dust, rather than gas, and that the total quantity of interstellar matter probably amounted to more mass than that of all the stars in the Milky Way combined. They had discovered that the galaxy was made of much more than stars alone.

Perhaps most important, Stebbins argued that his results could be explained by a very thin layer of obscuring matter extending through the entire disk of our galaxy (coinciding roughly with the visible extent of the Milky Way). Our solar system is immersed in this disk of obscuring matter, so we must look through more of it when we look along the Milky Way and through less of it when we look above and below the Milky Way. Stebbins liked to compare this model of the galaxy to a ham sandwich, in which the bread on the top and bottom represents the general distribution of stars, and the slice of ham in the middle represents the disk of interstellar matter running through the central plane. Further, and very important, if seen from outside, the disk of interstellar matter would appear like the dark layer of dust often seen in other galaxies when viewed edge-on. His results were exactly what we would expect if our galaxy was built like those other galaxies. This added strong support to the theory, still unsettled in the early 1930s, that our galaxy is comparable in structure and appearance to many others seen in the universe.

As Washburn astronomers continued their dedicated observing program of colors of type B stars, Albert Whitford's amplified photometer made it possible to reach dimmer stars. The Washburn team extended the program to still fainter stars at the much larger telescopes at California's Mount Wilson Observatory, which could reach stars farther south in the sky than could be seen from Madison. The result, in 1940, was the publication of a catalog of the colors of 1,332 type B stars, which widened the mapping of the interstellar matter across the sky and deeper into space.

By the late 1940s, Stebbins and Whitford had developed a very advanced photoelectric photometer using a new type of detector that offered much wider spectral sensitivity than

This edge-on view of the spiral galaxy known as NGC 891 clearly shows the layer of obscuring matter in the equatorial plane of the galaxy—what Joel Stebbins sometimes compared to the layer of ham in a ham sandwich. The less obscured and brighter layers of starlight above and below correspond to the bread in the sandwich. The B-star results of Stebbins and Huffer supported the idea that our own Milky Way Galaxy, with its layer of obscuring matter along the Milky Way, might look something like this if viewed from the outside. This supported the view that our own galaxy is similar in structure to other galaxies, rather than being a unique central galaxy in the universe, as some astronomers maintained. This image was made by Wisconsin astronomers using the WIYN 3.5-meter telescope at Kitt Peak National Observatory.
J.C. HOWK (NOTRE DAME), B. SAVAGE (WISCONSIN), N.A. SHARP (NSF) / WIYN / NOIRLAB / NSF

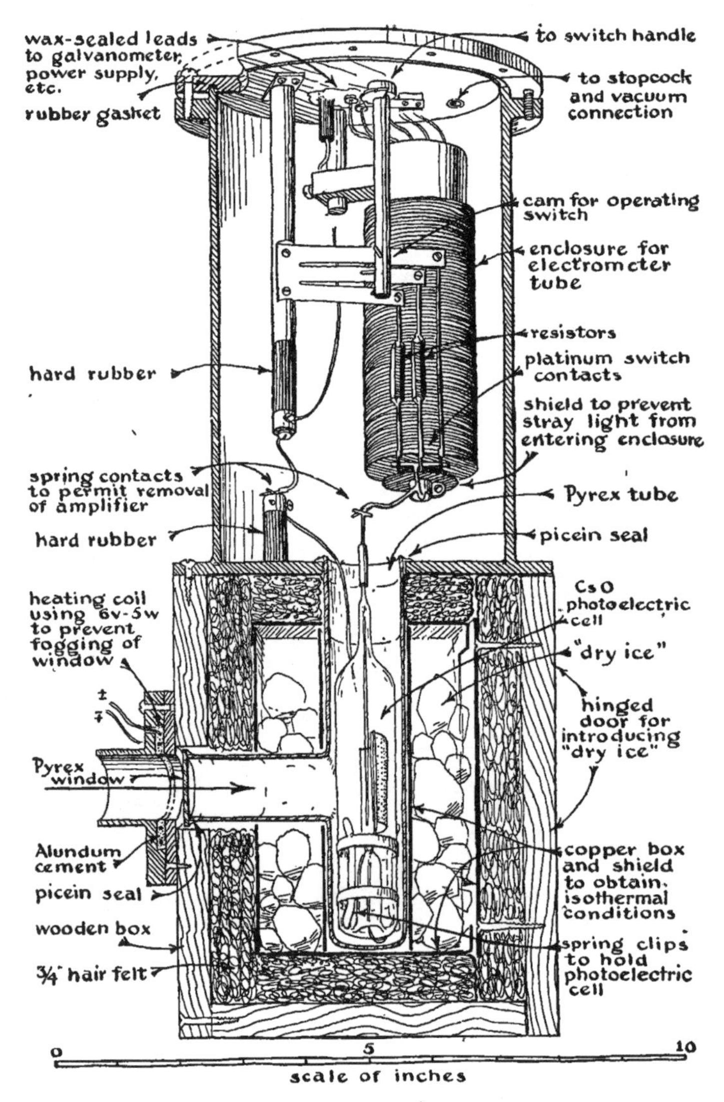

Fig. 11. Evacuated container for photocell and electrometer tube, with provision for refrigerating the cell to reduce thermionic emission.

This illustration of Whitford's six-color photoelectric photometer was published in 1938. The bottom half shows the cold box for holding the dry ice used to cool the detector tube (known as the "photoelectric cell"). Starlight from the telescope enters the Pyrex window from the left. The top half shows the brass tank containing the amplifier (or "electrometer") tube. The detector tube is enclosed in a glass extension of the tank so that it too operates in the vacuum. JOHN STRONG ET AL, *PROCEDURES IN EXPERIMENTAL PHYSICS* (1938)

the Kunz tubes. By cooling the detector with dry ice, the new photometer could detect light effectively across a very wide range of the spectrum, from the near infrared through the near ultraviolet, in six steps or "colors." Using their new six-color photometer on the large telescopes at Mount Wilson, the Wisconsin astronomers gathered enough color data on stars to understand the intensity of scattering in the interstellar medium as a function of the wavelength of light. In 1943, Stebbins and Whitford published their highly influential Law of Interstellar Reddening, which they would refine for years afterward, showing how the scattering of light by the interstellar medium increases as the wavelength of the light decreases. This understanding of the effects of interstellar matter provided a foundation for all subsequent studies of the extinction of starlight in our galaxy.

Albert Whitford's six-color photometer, pictured here, was used to establish the Stebbins-Whitford Law of Interstellar Reddening. The black wooden box held dry ice to cool the photodetector, which received light from the telescope through the window visible on the top surface of the box. Both the detector tube and the amplifier tube, inside the cylinder, operated in the vacuum maintained within the brass tank, very similar to the earlier Whitford amplified photometers. (Astronomer Ted Houck's coffee mug from the Orbital Astronomical Observatory is included here for scale.) JAMES LATTIS

27

SHRINKING THE MILKY WAY GALAXY

The story goes, according to Elizabeth Hart (Morse Huffer's wife, who occasionally worked at Washburn as a computer), that one day, possibly in late 1932, Joel Stebbins emerged from his office exclaiming, "We shrunk the universe!" He was using the word "universe" in an older sense, roughly equivalent to what we would call the galaxy, and specifically our Milky Way Galaxy. It's a fairly dramatic claim, so what did he mean?

Astronomer Harlow Shapley had become director of the Harvard College Observatory in 1921, just as Stebbins was transitioning from Illinois to Wisconsin. Shapley argued that the stellar system of which our Sun is a member—a vast disk of stars—effectively constituted the entire universe and was immersed in a halo of relatively small objects, primarily globular star clusters (spherical collections of hundreds of thousands of stars gravitationally bound to one another) and glowing, swirling clouds of gas, that is to say the spiral nebulae. Shapley's so-called super galaxy was, in his estimation, huge—perhaps two hundred thousand or even three hundred thousand light-years in diameter. He drew this conclusion from his measurements of distances to those globular clusters. In our sky we find that most of the globular clusters are seen in half of the sky roughly centered around the constellation Sagittarius. Shapley reasoned that because the globular clusters should be orbiting the center of mass of the galaxy, the crowding of globular clusters into that half of the sky showed that the Sun must be far from the center of the super galaxy, and instead (and here Shapley was right) the center of our galaxy must be off in the direction of Sagittarius.

Shapley had calculated the globular cluster distances from the Sun on the assumption that interstellar space was effectively transparent, which, if true, meant that he could relate the distance of a cluster to its relative brightness, because more distant clusters would be dimmer in a predictable way. But that only works if distance is the only factor dimming the light of a distant star cluster. Beginning about 1930, a few astronomers, including

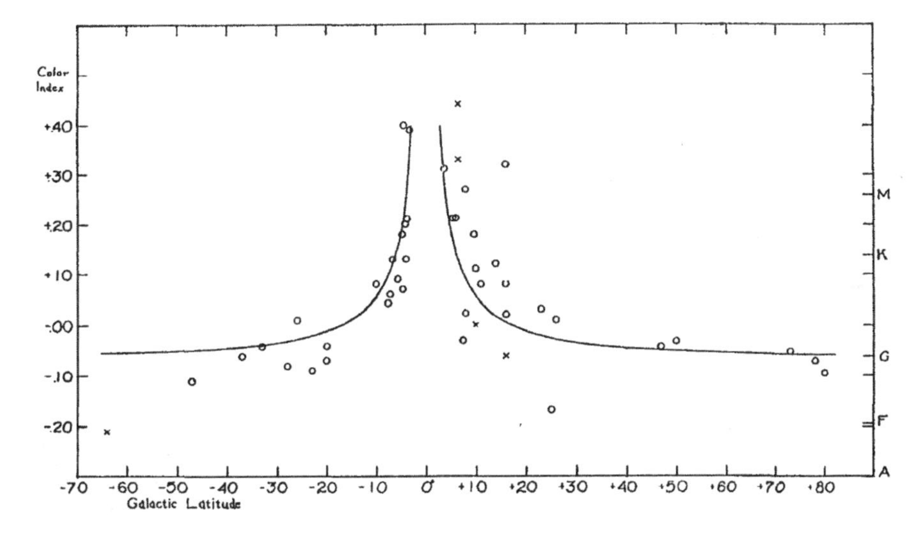

This diagram, illustrated by Joel Stebbins, shows how the reddening of globular clusters depends on the cluster's location in the sky. Each open circle represents the measured reddening of a globular cluster and its position relative to the plane of the Milky Way. The vertical axis plots the "color index" that Stebbins measured with his photoelectric photometers. A more positive color index (higher in the diagram) means the globular cluster is relatively redder, while a more negative number means it is relatively bluer. The horizontal axis plots galactic latitude, which is the angular distance above or below the plane of the Milky Way. As the diagram makes clear, globular clusters closer to the galactic plane (as latitudes approach zero) are dramatically redder than those located farther away. *CONTRIBUTIONS FROM THE MOUNT WILSON OBSERVATORY* (1933)

Stebbins, were finding evidence that interstellar space in our galaxy is far from perfectly transparent.

Using his methods of photoelectric photometry, Stebbins began measuring the colors of globular clusters to see whether those nearer the plane of the Milky Way are redder compared to other globular clusters far from the galactic plane. He found a strong effect. Globular clusters near the galactic plane are much redder than those at higher galactic latitudes, which means that their starlight reaches us after passing through a scattering medium, making the light redder and weaker, just as the setting Sun is reddened and weakened by its passage through Earth's atmosphere. This meant that Shapley's method was flawed.

The Washburn astronomers' program of measuring the reddening of hundreds of type B stars allowed them to estimate the effects of interstellar matter across much of the sky. Type B stars are far more numerous and more widely distributed than globular clusters. By using the type B star results to take into account the dimming of the globular clusters

This is a modern image of the globular cluster commonly known as Messier 71 (also called NGC 6838 or simply M 71). Found only about 4.5 degrees south of the galactic plane, M 71 is strongly reddened by scattering from interstellar dust. BOB FRANKE, FOCAL POINTE OBSERVATORY

In this edited image of M 71, the colors have been artificially adjusted to suggest how the star cluster would appear if it could be seen without the effects of interstellar reddening. BOB FRANKE, FOCAL POINTE OBSERVATORY

caused by interstellar matter, Stebbins was able to show that many of the most distant globular clusters were only about half as distant as Shapley had thought. This meant that the galaxy was not so "super," only about half as large as Shapley had claimed, hence Stebbins's exclamation about shrinking the "universe."

But the work on space reddening, a term Stebbins coined that is still used today, could be pressed a step further. Stebbins showed that the interstellar matter must actually be a large fraction of the mass of the galaxy, comparable or greater than the mass of all the stars put together, and that a large fraction of that matter must be dustlike particles in addition to gas. Thus, Stebbins's shrinking of Shapley's super galaxy was far more important than simply calculating an updated number for our galaxy's dimensions. His work advanced a new understanding of galaxies in several respects: that ours is typical rather than special; that fine dust dominates the medium spanning the space between the stars; and, as later astronomers would show, that this interstellar medium is key to understanding how stars and galaxies form, evolve, and interact. Stebbins's space reddening research took Wisconsin's exploration of the galaxy in a fundamentally new direction and constitutes a key contribution in our modern understanding of our universe.

28

SCANNING AND EXPANDING THE NEBULAE

Joel Stebbins and Albert Whitford spent the summer of 1933 at Mount Wilson Observatory "trying to contract the galaxy and expand some of the nebulae," as Stebbins put it in a letter to a colleague. In 1933, astronomers were still debating the nature of "extragalactic nebulae" (the term used at the time for what today we generally call galaxies). Mount Wilson astronomer Edwin Hubble had famously shown in 1924 that the Andromeda Nebula was actually a distant spiral cloud of stars rather than a nearby and much smaller spiral wisp of glowing gases, but astronomers remained unsure whether such spirals were actually galaxies like our own Milky Way, of comparable size and nature, or if they were perhaps much smaller star clouds around the edges of our super galaxy, as in Harlow Shapley's view. Put simply, were these spiral nebulae small objects relatively nearby, or were they large objects—what we would now call galaxies—relatively far away?

The relatively close and therefore best-studied extragalactic nebula is the famous Messier 31, or M31, also known as the Andromeda Nebula, located in the constellation Andromeda. Hubble had made a rather rough determination of M31's distance that put it well beyond the Milky Way Galaxy. At the distance Hubble calculated, the apparent size of M31 would make it a dwarf galaxy, not comparable in size to our own. Its apparent size in the sky was estimated from photographs. But what if photography could not detect the nebula's faint edges? Then it might actually be considerably larger, whatever its actual distance, and thus more like a peer of the Milky Way.

Like some other astronomers, Stebbins suspected that the spiral nebulae were more likely to be large, distant galaxies comparable to our own, in contrast to Shapley's opinion that they were dwarfish satellites of our super galaxy. If the "comparable galaxies" idea was right, then M31 should be larger than photographs seemed to show.

In the summer of 1933, Stebbins and Whitford decided to use the unique capabilities of their amplified photoelectric photometer, combined with the 100-inch telescope available

This modern image of the Andromeda Galaxy, also called Messier 31, is actually a mosaic of many smaller images. Modern imaging technology allows astronomers to see the outer regions of Andromeda's disk of stars that were undetectable with the photography of the early twentieth century. In the 1920s and early 1930s, some astronomers believed Andromeda was a small, nearby satellite of our own galaxy; today we know it's somewhat larger than our own Milky Way and about 2.5 million light-years away. M31 is our galaxy's closest neighbor of comparable size. ROBERT GENDLER

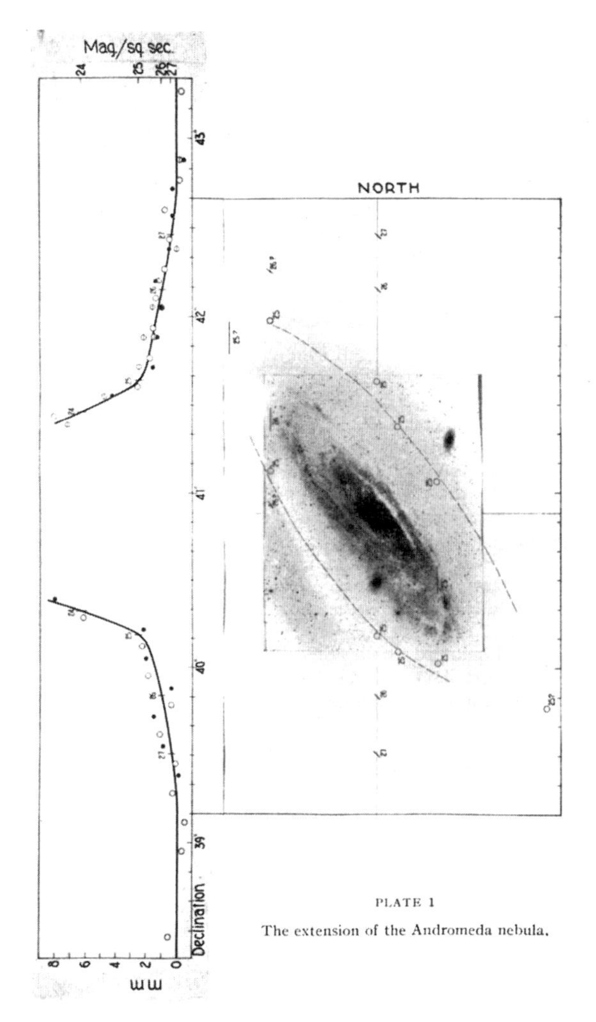

This diagram captures the dramatic results of Joel Stebbins's and Albert Whitford's work. The photometric traces along the left edge of the diagram represent the detectable starlight that exists well beyond the apparent edges of the Andromeda Nebula (which we now call the Andromeda Galaxy), as seen in the comparison photograph to the right of the traces. These results proved that photographic measurements had significantly underestimated that galaxy's size. Combined with the globular cluster and B-star results showing that our own galaxy is smaller than had been previously estimated, the work of Stebbins and Whitford supported the idea that our own galaxy is comparable in size to others, rather than being uniquely large.
CONTRIBUTIONS FROM THE MOUNT WILSON OBSERVATORY (1934)

to them at Mount Wilson Observatory, to measure the size of M31. Rather than having the telescope track the target, which was the usual procedure, their method was to keep the telescope stationary and allow Earth's rotation to carry M31 across the field of the photometer. The edges of the nebula would not be much brighter than the dark sky itself, which was the major reason that photographs could not be trusted to accurately reveal them. Instead, Stebbins and Whitford compared measurements of the nebula's surface brightness with offset measurements of the dark sky, so that the difference between the two measurements would indicate the light from the nebula alone. This was an adaptation of their differential methods of measuring variable stars.

Their results showed that the brightness profile of M31 extended well beyond the apparent edges of the photographic image. This was dramatic evidence that the size of the Andromeda Nebula, whatever its distance, had indeed been underestimated. The sizes of other spiral nebulae presumably had been underestimated as well. Wisconsin's photoelectric photometry did indeed "expand" the nebulae such that they seemed more likely to be comparable galaxies, though the case wasn't closed until some years later when more accurate distances were known. At the same that they were working to expand the nebulae, the astronomers were also working to "contract the galaxy" (or shrink the universe, as Stebbins would have put it) by studying globular clusters. Expanding the nebulae and shrinking the galaxy considerably reduced the apparent discrepancy in size that was implied by Shapley's super galaxy idea. Stebbins predicted, "When the size of our galaxy is as well-known as that of the Andromeda nebula, most of the inferred difference in scale between the two systems may disappear." The Wisconsin astronomers had offered powerful evidence leading to the recognition that our Milky Way Galaxy is actually pretty typical, just one among countless others.

29

WHEN IT'S DARK ON OBSERVATORY HILL

Observatory Hill's reputation as a romantic site was probably well established even before automobiles became common on campus in the late 1920s. The view from the summit of the hill, selected as the observatory's site by Governor Washburn himself, remains a lovely one, by day or night, and Washburn's astronomers certainly appreciated their location. Joel Stebbins even speculated that it might have been a factor in enticing James Watson to Wisconsin.

When the first lakeshore dormitories, Tripp and Adams, opened in 1926, suddenly hundreds of young men were crossing Observatory Hill each day to and from classes, often walking along the original Observatory Drive, which crossed the summit of the hill very close to the northern perimeter of the main observatory building (women, meanwhile, lived closer to town on the east side of campus). Washburn astronomers didn't mind the daytime foot traffic, but the cars that were increasingly arriving after dark along Observatory Drive were a different story. As the road became a well-known lovers' lane, it grew into a nuisance for the astronomers. The noise, lights, and even electrical interference from the cars' primitive ignition systems all disrupted astronomical observing. By 1932, a *Time* magazine article on Stebbins opened not with astronomy but with a wink at the hill's reputation: "University of Wisconsin jacks and jills like to go up Madison's Observatory Hill at night."

To the annoyance of the Washburn astronomers, their hilltop's romantic reputation soon eclipsed the astronomical one, attracting nationwide fame (and even a little notoriety) with the release of the popular song "It's Dark on Observatory Hill." The song was written for the Irving Berlin Publishing Company in 1934 by lyricist and former UW student Johnny Burke and composer Harold Spina. It was quickly recorded by several singers and orchestras, including Bob Crosby (the younger brother of Bing) and the Dorsey Brothers, and it remained a sentimental standard even thirty years after its original release in renditions by groups such as the Ray Conniff Singers and the Lettermen.

"It's Dark on Observatory Hill" was first published as sheet music by Irving Berlin's company in 1934. The song was recorded by the Dorsey Brothers orchestra that same year and by several other singers and ensembles over the next three decades.
IRVING BERLIN, INC.

For many years throughout the 1930s, Stebbins packed up his photometer equipment and drove to California for summer observing at Mount Wilson. He would later recall how the song's popularity followed him: "Once when I was checking my car across the state line into California the inspector noticed the license plate and said, 'Wisconsin—that is where it is dark on Observatory Hill!' To which I replied, 'You're telling me!'" But it wasn't always a joke. In the spring of 1935, Stebbins wrote to university president Glenn Frank in some detail, with specific dates and the number of cars containing "necking couples," and requested action to restrict traffic on the road. The problem peaked in the summer of 1935, with more cars (Stebbins was counting, as a good scientist would) and even greater disruptions to the astronomers' work. Stebbins wrote again to President Frank that fall, insisting on an investigation. Finally, and to the relief of the astronomers, traffic was rerouted by 1937 to a new road, the modern Observatory Drive, which kept cars away from the hill's summit and the observatory building. The road to the top of the hill was barricaded each night after dark, prompting Stebbins to remark dryly, "I don't know how much trouble the barricade on the road is making for the watchman, but the Hill has been transformed into a convenient place to work."

SPOTLIGHT

WASHBURN OBSERVATORY STAFF, 1936

The mid-1930s were a time of great change, fame, and triumph for Wisconsin astronomers. Though the occasion and photographer are unknown, this spring 1936 portrait taken on the eastern steps of Washburn Observatory immortalizes a team of astronomers who were unique at the time in their power to explore the galaxy.

The staff of Washburn Observatory, then the world's leading center for astronomical photoelectric photometry, both in the development of instruments and techniques as well as the scientific research and results that followed. DEPARTMENT OF ASTRONOMY, UNIVERSITY OF WISCONSIN–MADISON

In the back row, left to right, are Edwin J. Bernet, a student assistant working on the space reddening program; Gerald E. Kron, an engineering student and observatory assistant who was about to embark on his own astronomical career as a graduate student at Lick Observatory; and Albert Whitford, then a research associate at Washburn. It's possible this photograph commemorates Kron's imminent departure from Wisconsin. Around the time the photo was taken, Whitford and Kron were involved in experiments with a prototype of the newly invented electron multiplier photodetector (also called the photomultiplier tube, or PMT), descendants of which would become the defining technology of their postwar research careers.

In the front row, left to right, are senior astronomer Morse Huffer; Elsie De Noyer, the observatory's secretary, stenographer, and receptionist; and Washburn director Joel Stebbins. With Stebbins's guidance, Huffer was in the midst of measuring the colors of well over a thousand stars to probe the nature and extent of interstellar material in the disk of the Milky Way Galaxy. In mapping the extent and effects of the interstellar medium, they overturned earlier models of our galaxy, provided new evidence that it is one of many comparable galaxies, and revealed that a galaxy is made of far more than stars alone. The climactic results of their work were just being sent off for publication in early 1936.

Invisible here are the many other essential women, like Elizabeth Huffer, who worked as computers, carrying out the repetitive data reduction that transformed the handwritten observations and measurements in an observer's notebook into the scientific results that ended up in astronomical journals. Also invisible are the many other student night assistants who, like Bernet, worked with a senior astronomer, usually Huffer, to accumulate the seemingly countless observations that the human computers distilled into the color measurements of Milky Way stars bright enough to serve as probes of our galaxy's interstellar clouds.

This small team encompassed all the functions of a modern research project: questions about structures and patterns in nature provoke ideas and theories. The investigation and testing of those ideas and theories motivates new observations, methods, and technologies for more sensitive measurements. Those measurements help scientists test and decide between competing theories. And the cycle continues as the new results and new understandings inevitably create still more questions about ever more subtle patterns, structures, and effects that we encounter in our explorations of the universe.

30

A HOME FOR WASHBURN'S DIRECTORS

THE OBSERVATORY RESIDENCE

Today, the oldest building on Observatory Hill houses the University of Wisconsin's Robert M. La Follette School of Public Affairs, but for its first seventy years it was the home of the directors of Washburn Observatory and their families. The home was built in 1855 by Madison resident Loring Guild on what was then Guild's private property, and it passed into university hands shortly after the legislative reorganization of the university in 1866, when Dane County purchased the building along with nearly two hundred acres along Lake Mendota to be used for the university's Experimental Farm. For a while the building was the home of university professor Daniel Read, but around 1867 it was remodeled and became the official residence of the university president. The last of the presidents to enjoy the scenic beauty of the site was John Bascom, who lived there until the regents relocated the president to a new house on State Street more convenient to the campus and State Capitol. By agreement between former governor Cadwallader Washburn and the Board of Regents, the building on Observatory Hill was made available to house the director of the new Washburn Observatory.

When James Watson, the observatory's first director, moved to Madison from Ann Arbor in June 1879, Bascom was still occupying the Observatory Hill home because his new house wasn't yet ready, so Watson and his wife stayed with Washburn at his Edgewood estate. It wasn't until early September 1880 that Bascom vacated Observatory Hill so the Watsons could relocate from Edgewood to the hilltop. But the Watsons lived there only briefly. James Watson died suddenly that November, and his widow promptly moved back to Ann Arbor.

Edward Holden arrived from Washington, DC, in March 1881 and lived in the Observatory Hill residence with his wife and children and at least one student observatory

From 1880 to 1948, the Observatory Residence housed the observatory directors and their families. The home offered a fine view of Lake Mendota and originally included a barn, seen here behind the tree. In the distance, to the right of the barn, the building and dome of the Student Observatory can be seen.
WHI IMAGE ID 24030

assistant, Emma Gattiker, who had a room in the house for a while. In the summer of 1881, Holden arranged for the first photo of the observatory to be taken from the roof of the Observatory Residence.

After Holden left Madison in November 1885, no one occupied the residence until August 1887, when newly appointed director George Comstock moved in. Comstock, a bachelor, may have housed some of the observatory assistants in the large house, but it became a family home once again when Comstock married Esther Everett of Madison in June 1894. Comstock served as director until 1922, and his daughter, Mary, born in December 1895, was the only child born and brought up on Observatory Hill during the astronomers' time there.

The fourth and final director to live in the Observatory Residence was Joel Stebbins, who moved to Madison in June 1922 with his wife, May, and their children, fifteen-year-

Every winter for many years, students transformed Observatory Hill into a legendary toboggan run, from the top of the hill out onto the frozen surface of Lake Mendota. This view from the lake, in a photograph taken in 1911, shows the astronomical complex atop the hill. From left is the Observatory Residence, the barn, the Student Observatory, and Washburn Observatory itself. UNIVERSITY OF WISCONSIN–MADISON ARCHIVES, MADISON, WISCONSIN

old Robert and twelve-year-old Isabelle. Joel and May Stebbins were enthusiastic hosts, regularly using the Observatory Residence for social gatherings and visitors, including not only visiting astronomers but also international peace activists such as John Herman Randall and major figures of the international Bahá'í community such as Ruhi Afnan and Martha Root. In 1929, May Stebbins hosted at the residence a celebration of the ratification of the Kellogg-Briand Pact, an international peace agreement that aspired to outlaw war.

Of course, like everyone else, the Stebbins family faced the challenges of the times, such as a quarantine in February 1923 when May, Robert, and the family maid all came down with scarlet fever. Joel and Isabelle escaped both the disease and quarantine by taking refuge in the observatory itself, which was equipped with a small kitchen and lodgings that in earlier days had been occupied by student assistants, including Stebbins himself during his student days of 1900–1901.

A familiar figure on Observatory Hill in the mid-1920s was Tycho, the Stebbins family dog. Tycho posed cooperatively in front of the observatory in a group photo during the American Astronomical Society meeting in 1927, and he received special greetings later in a letter to Stebbins from Japanese astronomer Issei Yamamoto, who had attended the meeting. But Tycho, who was allowed to roam the hill, also seems to have been the source of mischief around the university, leading to a letter Stebbins received from the superintendent of buildings and grounds that reported a complaint about the dog and expressed concern that Tycho might be a danger to the many visitors on the hill.

Throughout Stebbins's time in the Observatory Residence, the social highlight may have been Isabelle's wedding, held at the residence on June 15, 1935. Her husband was Theodore "Ted" Dodge, a Madison native and son of UW English professor Robert Elkin Neil Dodge. For the large crowd assembled, as the local newspaper later reported, astronomer Morse Huffer, an accomplished musician, played the piano, including Franz Schubert's "Ave Maria," Ethelbert Nevin's "The Rosary," and Claude Debussy's "Clair de Lune."

Joel Stebbins retired in 1948 and moved out of the Observatory Residence. He and May bought a house in Madison but soon relocated to California. The next observatory director, Albert Whitford, had already settled with his family elsewhere in Madison and didn't need to move into the residence. It was then designated for other uses, including as the Observatory Hill Office Building, before becoming the home of the La Follette School, as it remains today.

31

IN THE STEW

CONTROVERSY AND THE STEBBINS-WHITFORD EFFECT

In the 1940s, the colors of galaxies and something known as the Hubble Law led Washburn astronomers into the jaws of one of the great controversies of mid-twentieth century scientific cosmology: the debate between advocates of the Big Bang theory, who argued that the universe had begun at some finite time in the past, after which it had expanded and cooled; and their opponents, advocates of the Steady State theory, who argued that the universe is immeasurably old and exists in a steady state that looks the same, on the average, everywhere in space and time.

In 1912, astronomer Vesto Slipher of Arizona's Lowell Observatory began using spectroscopic observations to measure the velocities of spiral nebulae (what we now call galaxies) moving toward or away from us, which astronomers call radial velocities. Slipher soon found that with very few exceptions, these galaxies are moving away from us. In 1924, Edwin Hubble showed that these spiral nebulae were at distances far beyond our Milky Way Galaxy and must be galaxies in their own right. Hubble succeeded in measuring the distances to a number of galaxies, and by 1929 he was able to show, using Slipher's radial velocity data, that the more distant a galaxy, the faster it is receding from us, revealing what has become known as the Hubble Law.

Joel Stebbins and Albert Whitford had used their work on the colors of globular clusters and the reddening of type B stars to improve scientific understanding of the size and structure of our own Milky Way Galaxy, as well as the nature of other comparable galaxies (or "extragalactic nebulae"). Beginning around 1940, Stebbins and Whitford took their expertise in measuring the colors of faint astronomical objects and applied it to ever dimmer extragalactic nebulae, by using the Wisconsin six-color photometer (the same instrument they used to establish their Law of Interstellar Reddening) on the large telescopes at

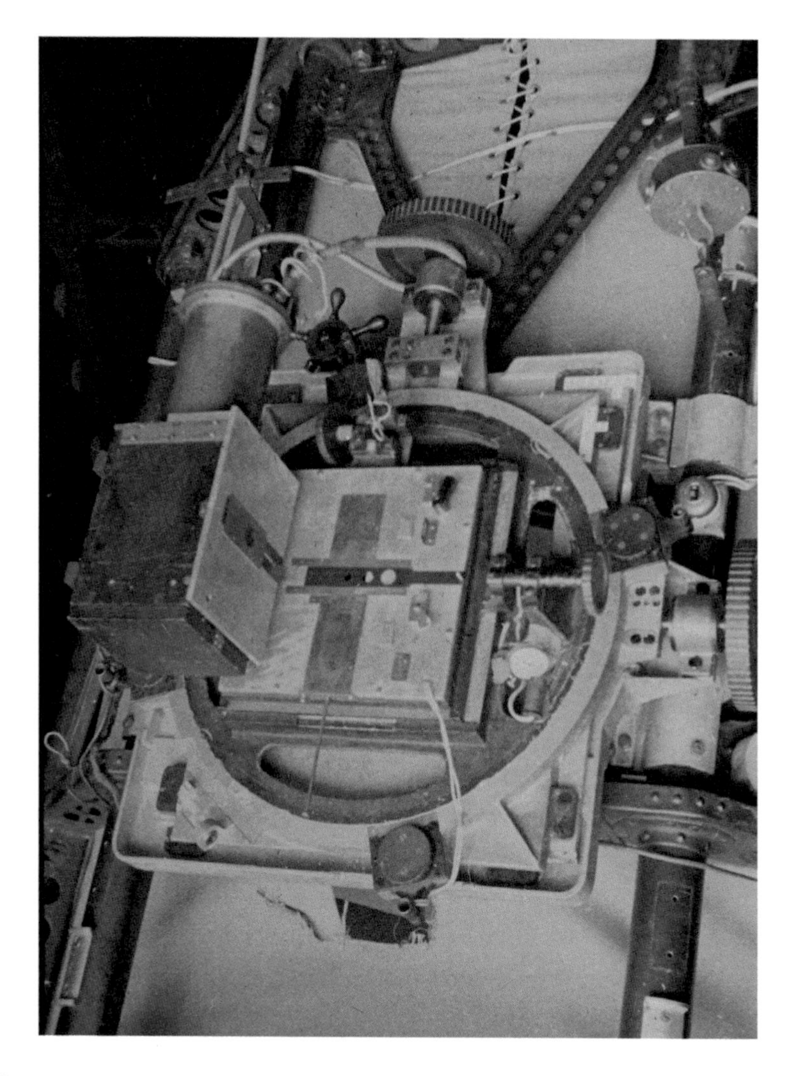

In 1941, the Wisconsin six-color photometer was mounted on the Mount Wilson Observatory 60-inch reflector. (The same photometer is pictured in chapter 26.) Somewhat challenging to see here, the photometer is roughly centered vertically and appears as a black, rectangular box attached to a brass cylinder. The photometer unit is hinged away from the focal plane of the telescope to show the selectable apertures, on the telescope side, and the instrument aperture, on the side of the box. It would be rotated into position to make observations. The black box is a wooden chamber that was loaded with dry ice daily to keep the detector tube cold for infrared measurements. The brass cylinder is the vacuum tank containing the amplifier tube and, by a glass extension into the cold box, the light detector tube. Though Whitford was already at MIT for "war work," he may have gotten leave to help Stebbins and research assistant Gerlacus Bouricius begin their summer observing work in 1940 and 1941. The interruptions of World War II prevented the completion of their observations at Mount Wilson until the summer of 1947. DEPARTMENT OF ASTRONOMY, UNIVERSITY OF WISCONSIN–MADISON

This is one of the galaxy fields, in the constellation Boötes, that Joel Stebbins and Albert Whitford used for their six-color studies that led to the Stebbins-Whitford Effect. The target galaxy itself, along the top-center edge, is labeled "1" with an arrow. The small circle around an unrelated object shows the apparent size of the photometer's aperture on the sky.
ASTROPHYSICAL JOURNAL (1948)

Mount Wilson Observatory. Their goal was to investigate whether the color of a distant galaxy could be used as a proxy for the actual spectral redshift, which is what Slipher used. Color measurements of faint objects were much easier to make than spectral measurements of the redshift, so the six-color technique, if it worked, would be a big advantage.

In 1948, Stebbins and Whitford published their first results on the colors of extragalactic nebulae. As expected, more distant galaxies were redder in color but, surprisingly, not exactly as predicted by the Hubble Law. After correcting for the Hubble redshift, a certain category of galaxies, the giant ellipticals, seemed to be systematically redder than expected, and the discrepancy increased with distance. This became known as the Stebbins-Whitford Effect.

This finding was a big problem for the Steady State theory. As we look farther out into space, we see further back in time, so distant galaxies are showing us the universe as it used to be, millions and even billions of years ago. According to the Steady State theory, the universe of the distant past should generally look, on average, like the universe today. Yet Stebbins and Whitford seemed to have found that giant elliptical galaxies were systematically redder when the universe was younger. This in turn implied a general evolution of the universe, contradicting the Steady State theory and potentially corroborating the Big Bang theory. So it's no surprise that the Stebbins-Whitford Effect quickly attracted strong criticism from Steady State advocates. But the actual downfall of the Stebbins-Whitford Effect eventually came from much closer to home.

Young astronomer Arthur D. Code had joined the staff of Washburn Observatory in 1951, by which point Stebbins had retired and Whitford had become director. Code was an instrumentalist and observer, so he learned the Stebbins-Whitford methods of photoelectric photometry backwards and forwards. In studying their work on their "effect," Code noticed a flaw that had escaped his senior colleagues. Essentially, Stebbins and Whitford had calculated how the color of a red-shifted galaxy should be expected to appear based on assumptions about how much light would be emitted by galaxies in the far blue and ultraviolet parts of their spectra. But Code realized that their assumptions could be wrong, since in those days the spectra of stars and galaxies, especially in the ultraviolet, were poorly understood. It was possible, Code reasoned, that the apparent excess reddening of those distant galaxies was a result of those potentially flawed assumptions about the light output of galaxies. Whitford recognized the validity of Code's critique, and by 1956 he published a retraction of the Stebbins-Whitford Effect. Subsequent research on detailed spectra of galaxies showed Code's critique to be correct.

In 1958, reflecting philosophically on the episode, Stebbins (who by then had been retired for a decade), observed that being wrong is sometimes simply part of advancing science: "As long as you are doing research you can expect to be in difficulties; you might as well become reconciled to being in a stew for the remainder of your career."

In retrospect, it is no surprise that the resolution of the Stebbins-Whitford Effect came from Code, who, along with other Wisconsin astronomers, would go on to pioneer the instruments and techniques of ultraviolet photometry in the early Space Age. The understanding of how the spectra of stars and galaxies actually behave in the ultraviolet part of the spectrum was one of the primary motivations of Wisconsin's early space astronomy efforts and provided the solution to the mystery that had bedeviled the earthbound efforts of Stebbins and Whitford. As Code and Whitford showed us, good scientists must always be critics of their own theories.

32

THE COUNTY TELESCOPE ARRIVES

PINE BLUFF OBSERVATORY

The development of mountaintop observatories in North America began in the 1880s with the opening of Lick Observatory on Mount Hamilton, outside of San Francisco. As research telescopes and observatories grew larger and more expensive, the importance of optimizing their observing conditions quickly began to outweigh the inconvenience of building atop remote mountains, and by the early twentieth century, the westward movement of astronomers to mountaintop observatories was well underway. In Madison, Washburn Observatory director George Comstock wanted to move westward, too—but only farther west along the shore of Madison's Lake Mendota. Comstock urged university regents to acquire some land there in order to escape the westward development of the campus that would eventually engulf the once dark and isolated Washburn Observatory.

Though the university did eventually acquire much of that land, Comstock ultimately decided against moving the observatory to what is now the neighborhood of Eagle Heights, in part because of the now quaint reason that the city streetcar lines did not run that far westward. But Comstock also wanted an updated telescope and was reluctant to pursue any plan that would simply relocate the old one, so he seems to have opted to postpone planning for a move unless expensive new instruments were part of it. Perhaps it's just as well: had his plan moved forward, any relief from urban light and smoke would have been temporary.

As director, Joel Stebbins also angled for darker skies without having to go as far as California. Stebbins scouted potential sites for a "country telescope" on Saturday afternoon hikes in the company of his friend, physics professor Leonard Ingersoll, whose wife, Helen, was the daughter of Washburn Observatory astronomer Albert Flint. Initially, Stebbins was partial to a site on a peninsula jutting into Lake Kegonsa, southeast of Madison. But he admitted that an alternative site suggested by Albert Whitford, fifteen miles west of campus

The opening of Pine Bluff Observatory on June 30, 1958, brought together, from left to right, incoming observatory director Arthur Code; outgoing director Albert Whitford; and Whitford's predecessor, Joel Stebbins. University president Conrad Elvehjem is pictured second from right. UNIVERSITY OF WISCONSIN-MADISON ARCHIVES, MADISON, WISCONSIN

on a hill above the little town of Pine Bluff, was the astronomically superior location, and it was already owned by the university.

Stebbins set in motion the plans to transition Washburn's work to a new location, but he retired in 1948, before any detailed plans could advance. The task of funding, designing, and building the new observatory fell to Whitford, in his new role as Washburn's director, and university president Edwin B. Fred, who acquired nearly all the support for the project from the Wisconsin Alumni Research Foundation. Whitford knew that the Yerkes Observatory optical shop was planning to make a 36-inch reflecting telescope for photometry destined to operate at the McDonald Observatory in Texas (which Yerkes operated in those days). So, he arranged a deal in which a second identical telescope would be made at the same time, resulting in a discount for both observatories. The new telescope would, of course, be well suited for that Wisconsin specialty, photoelectric photometry.

The new Pine Bluff Observatory opened on June 30, 1958, an event marked by the one-hundredth meeting of the American Astronomical Society, whose members gathered

The main building at Pine Bluff Observatory houses the 36-inch reflecting telescope in the dome on its upper level. The ground level includes living quarters for an astronomer, as well as a darkroom for developing the photographic plates on which spectra were recorded. The basement level contained the control room for the telescope and instruments, as well as mechanical and electronics shops. DEPARTMENT OF ASTRONOMY, UNIVERSITY OF WISCONSIN-MADISON

in Madison for the conference and attended a picnic and dedication ceremony at Pine Bluff. The opening day of the new observatory was the last official act Albert Whitford performed as director of Washburn Observatory before he left, like Edward Holden, to become director of Lick Observatory. The next director of Washburn Observatory, Arthur Code, would take up his duties the following day. Joining Whitford and Code at the ceremony were new university president Conrad Elvehjem, other university dignitaries, and some four hundred guests from around the world. Stebbins returned to Madison to give the dedicatory address, which included an informative and idiosyncratic account of the history of Wisconsin astronomy. Stebbins concluded his remarks with a sly reference to the old song: "Of the nights to come, we trust that many will be both clear and dark on the new Observatory Hill."

The new observatory, familiarly called PBO, was indeed the astronomical workhorse that Stebbins and Whitford had hoped for. Some work remained familiar, such as the photometric monitoring of eclipsing variable stars, now moved from the Washburn refractor to the new 36-inch reflecting telescope. But much was innovative, such as the first photometric measurements of ionization levels in diffuse nebulae, taken by university astronomer Donald Osterbrock, which provided estimates of the invisible ultraviolet flux of hot stars that a few years later would be directly measured by Wisconsin photometers in space. Osterbrock's specially made telescope incorporated parts thriftily cannibalized from the old Repsold meridian circle.

PBO also became the site of the world's first robotic observatory, called the Automatic Photometric Telescope, built in 1965 in part as a prototype for testing the instruments that would operate in Earth's orbit on NASA's Orbiting Astronomical Observatory (OAO) satellites. Astronomer Theodore "Ted" Houck, who had earned his PhD while working for Whitford at Washburn Observatory, was motivated by the idea that the robotic telescope would make good use of Wisconsin's scarce astronomically perfect nights, often called photometric nights. Astronomical systems engineer John McNall did most of the electronic and software design, and together he and Houck made the automatic telescope work.

The telescope itself was the same design as the eight-inch off-axis stellar photometric telescopes being made by the Space Astronomy Laboratory for the OAO project, including its then-innovative digital control system. The robotic observatory monitored sky conditions and, when permissible, opened the roof of its small shed enclosure to make and record photometric observations of a list of target stars until dawn, unless it detected cloudy skies that required closing up and shutting down. It could operate for several days at a time without human control.

The optical design of PBO's 36-inch telescope was optimized for its intended work of photometry and spectroscopy. It was the first new research telescope acquired by Wisconsin astronomers since the purchase of Sherburne Burnham's six-inch telescope in 1882, and it was by far the largest under the control of Wisconsin astronomers until the 3.5-meter WIYN telescope in Arizona came along nearly forty years later. The instrument mounted at the bottom of the telescope is, of course, a Wisconsin photometer—a third-generation instrument using a photomultiplier tube for a detector. The astronomer could look through the telescope field by means of a mirror to select the target star; the astronomer moved the mirror to direct starlight into the photometer. Results were recorded as tracings on paper by the recorder in the rack on the far left. UNIVERSITY OF WISCONSIN–MADISON ARCHIVES, MADISON, WISCONSIN

The world's first robotic telescope was mounted at Pine Bluff Observatory in a roll-off roof shed, shown here opened to the sky, northwest of the main PBO building. The eight-inch off-axis telescope, its attached photoelectric photometer, and its mounting/pointing system were controlled from the main PBO building by a DEC PDP-8 computer. UNIVERSITY OF WISCONSIN–MADISON ARCHIVES, MADISON, WISCONSIN

In addition to many years of stellar photometry and high-resolution spectroscopy, the PBO 36-inch reflector was used in two space shuttle missions as the Earth-based counterpart of a space instrument, the Wisconsin Ultraviolet Photo-Polarimeter Experiment (or WUPPE), which made ultraviolet observations of the polarization of light from stars. The PBO telescope's job was to observe the same targets and measure the polarization in the visual light that can reach Earth's surface in order to complement the orbital ultraviolet data.

As much larger telescopes in Arizona and South Africa have become available to Wisconsin astronomers, and as light and electrical pollution from surrounding development has degraded viewing conditions, PBO is now rarely a primary site for scientific observations. But it has been the development platform for several innovative astronomical instruments, where a moderately dark sky paired with quick connections to Madison's laboratories and shops makes an efficient combination. A good example is the Wisconsin Hydrogen-Alpha Mapper (WHAM), which has surveyed the hydrogen atoms that surround and constitute much of the mass of our Milky Way Galaxy. After being developed at PBO by Wisconsin astronomer Ron Reynolds and his team, WHAM moved on to complete surveys of the visible galaxy from Kitt Peak National Observatory in Arizona, and then Cerro Tololo in Chile, for access to the southern sky.

Astronomical engineer and computer specialist John McNall (left) and an unidentified colleague work in the basement control room of PBO, where both the 36-inch telescope and the robotic telescope were controlled.
DEPARTMENT OF ASTRONOMY, UNIVERSITY OF WISCONSIN–MADISON

33

THE ASTRONOMERS LEAVE OBSERVATORY HILL

After serving as the university's astronomical hub for eighty years, by 1960, Observatory Hill had become an astronomical afterthought. When May and Joel Stebbins moved away in 1948, the director's residence was converted to university office space. But Washburn Observatory itself was still the home of the Astronomy Department until the summer of 1958, when astronomical observing moved largely to the new Pine Bluff Observatory. At that point, the Washburn 15.6-inch telescope entered its long retirement, primarily serving astronomy classes and the general public. In the spring of 1959, university astronomers began moving their offices from the old observatory building to the top floor of the recently completed east wing of Sterling Hall. By July 1960, the Student Observatory was physically relocated to nearby Fitchburg, to serve as the club observatory of the Madison Astronomical Society. In the space of two years, astronomy had largely disappeared from Observatory Hill, except for public visitor nights on the first and third Wednesday evenings of each month, as long as the sky was clear.

In their new quarters in Sterling Hall, the astronomers had far more office, library, shop, and classroom space, not to mention closer proximity to their Physics Department colleagues. Their unique space included the roof of Sterling Hall's east wing, which sported four domes. The largest dome housed the planetarium, which was moved from its inconvenient location in the old university machine shop building (on the site where Helen C. White Hall now stands), where it had served to teach astronomy and celestial navigation during World War II and after. The other domes were designed to house telescopes. During the move out of Washburn Observatory, Burnham's famous six-inch Clark refractor was out of town, as Albert Whitford had lent it to advanced amateur astronomer Jerome J. Knuijt, who had it mounted in his own observatory in Appleton, Wisconsin. But by late 1959, the Burnham telescope was returned to Madison to be mounted in one of the Sterling Hall domes where it, the university's oldest telescope, continues in service today.

Four rooftop domes were built on Sterling Hall specifically for the Astronomy Department. The largest dome, on the far right, houses the planetarium. The dome second from right contains the venerable Burnham telescope. The other two domes have housed a succession of telescopes serving the changing needs of student research and instruction. JEFF MILLER, UNIVERSITY COMMUNICATIONS

The bombing of Sterling Hall in August 1970 traumatized the community, killed physicist Robert Fassnacht, and inflicted tremendous structural damage. The building was nearly gutted by the force of the explosion, right up to the offices of astronomers on the sixth floor. Concrete piers supporting the rooftop telescopes helped prevent the total collapse of the east wing. WHI IMAGE ID 33885

The telescope mountings on the roof of Sterling Hall may have played an important, if unintended, role in one of the most traumatic events in the university's history. Early in the morning of August 24, 1970, four domestic terrorists exploded a truck bomb at the base of Sterling Hall's east wing, killing one physicist, injuring three others, and doing enormous damage to the building. According to Arthur Code's estimation, the building itself might have collapsed if not for the extra support of the three sturdy concrete piers running vertically through the building from the telescope domes to the foundation. Even on the top floors of the building, the Astronomy Department was reduced to chaos for months, both physically and academically, as works in progress had been blown out of windows and scattered across campus.

By the fall of 1959, the old Washburn Observatory building housed the newly formed Institute for Research in the Humanities, where, under the directorship of professor and historian of science Marshall Claggett, it supported advanced scholarship in history, classics, philosophy, and other humanistic fields for the next half century. In the tranquility atop Observatory Hill, Washburn's original 15.6-inch telescope stood among the institute's scholars as testimony to the building's astronomical past.

PART THREE

"When you come to a branch in the road, go both ways."

A frequently shared witticism by Arthur Code, the sixth director of Washburn Observatory and recipient of the 1992 NASA Distinguished Service Medal

34

THREADING THE NEEDLE

FROM CLASSICAL ASTRONOMY TO ASTROPHYSICS

How did the venerable Washburn Observatory navigate the tidal shift away from the "old" astronomy to emerge as a leader in the new age of astrophysics? Modern astrophysics arose late in the nineteenth and early twentieth centuries based on a "tripod" of technologies: photography, spectroscopy, and photometry. Spectroscopy allowed scientists to study light by splitting it into various wavelengths to produce a spectrum that could then be analyzed. With photometry, scientists could precisely measure the relative brightness of stars and other celestial light sources. Both technologies helped astronomers glean rich and essential information about the universe, such as a light source's orbital motions, distance, or chemical composition. And both were transformed by the use of photography.

Earlier generations of astronomers had primarily used their own eyes to look through their telescopes in order to measure positions, make timing determinations, and describe the visible features of planets, comets, the Moon, and the Sun. But telescopes and other instruments designed for the human eye generally were not well suited to photography. The emergence of photography transformed some activities so completely, such as measuring star positions, that entire categories of instruments became obsolete—meridian circles, for example. Astronomical photography, photometry, and spectroscopy (aside from the solar kinds) need as much light as possible, and that led to ever larger telescopes, particularly driving the rapid development of large reflecting telescopes.

The same observatory that had been well equipped in the 1880s was antiquated by the 1920s. Decades of classical double star measurements using Washburn's 15.6-inch telescope, as well as position measurements at the meridian circle, came to an end with the retirements

Early generations of Washburn astronomers viewed celestial objects with their own eyes through the eyepiece of the telescope. DEPARTMENT OF ASTRONOMY, UNIVERSITY OF WISCONSIN–MADISON

Washburn Observatory was one of the earliest observatories in the world to have an electronics shop. Here, the first amplified photometer, whose brass vacuum tank is the cylinder on the right, is set up for bench testing ca. 1932. Albert Whitford's soldering iron, visible at bottom center, was unfamiliar to most astronomers and became his trademark tool in his early days doing Wisconsin photometry at Mount Wilson Observatory. DEPARTMENT OF ASTRONOMY, UNIVERSITY OF WISCONSIN–MADISON

of Albert Flint in 1920 and George Comstock in 1922. Within just a few years, by 1925, Washburn had transformed into a high-tech "electrical" observatory, with banks of batteries, wires, meters, air hoses, and bubbling acid for drying the air. It would have fit perfectly into a Tom Swift story of those days.

George Comstock, the consummate classical astronomer, hand-picked Joel Stebbins, the pioneer of photoelectric photometry, as his successor. As director of the University of Illinois Observatory, Stebbins had grown dissatisfied with the support and recognition he received in his nearly twenty years there. UW president Edward Birge promised Stebbins significant research funding in Madison and a free hand in reshaping the work of the observatory. Compared to Urbana, Madison offered a larger telescope at a superior site, not to mention the Observatory Residence and the spectacular vistas from the hill that came with the job. (Stebbins was known to joke about the contrast between Washburn, overlooking Lake Mendota, and the Illinois Observatory, where he had a view of a cornfield.)

In his turn, Stebbins ensured he was succeeded by Albert Whitford as Washburn Observatory's fifth director, thus guaranteeing the continuity of the productive research

Albert Whitford poses at the operating station of the Washburn Observatory amplified photometer, which he invented and put into operation in early 1932. The vacuum tank of the photometer is mounted at the end of the telescope, where the Kunz cell measures the starlight, and the amplified signal runs to a galvanometer. The galvanometer's response is read by the astronomer on the horizontal glass scale, held in a lab stand visible just to the right of Whitford's control box. Typical of the new age of astrophysics, the astronomer's acute vision is no longer essential to the measurement. WHI IMAGE ID 17477

programs they had developed over the previous quarter century. Whitford, who had entered astronomy with a PhD in physics, embodied the deep familiarity with spectroscopy, photometry, and electronic detectors that characterize modern astrophysics. When the University of Wisconsin hired Arthur Code away from the California Institute of Technology in 1958 to be the sixth director, it marked the return of a familiar face, as Code had previously worked at Washburn under Whitford from 1951 to 1956. As was the case when attracting Stebbins from Illinois, the university promised Code free rein and research support to pursue his vision of creating space-based telescopes and other instruments—an area in which Wisconsin astronomers quickly excelled, in part owing to the adaptability of electronic instruments (compared to photography, which was the defining technology elsewhere) to space applications.

Looking back, every director of Washburn Observatory was an enthusiast and innovator in astronomical technology, though the specific technologies changed. Watson designed the novel, if peculiar, Solar Observatory for his Vulcan telescope. Holden encouraged reflecting telescopes and photography (doing this important work as director of Lick Observatory). Comstock developed new methods of visual astrometry, adapting to his purposes the Loewy device and diffraction effects. Stebbins pioneered the first generation of electronics in astronomy, while Whitford invented the second generation and then expanded applications of the third by adapting the earliest photomultiplier detectors. Finally, Code launched Wisconsin photometry into space.

Stebbins liked to emphasize the importance and effectiveness of developments at the "small end" of telescopes—for example, pointing out that Whitford's amplifier, by extending measurements to much fainter stars, was equivalent to doubling the aperture of the telescope, but orders of magnitude cheaper. This leverage of working at the "small end" explains how the Washburn 15.6-inch refractor continued to be a productive research instrument, used by Morse Huffer to monitor eclipsing variable stars, well into the 1950s, by which time, in any other hands, it would have been an antique. It took the completion of Pine Bluff Observatory to drive Washburn's refractor, by then quite unusual in its prolonged utility for research, into active retirement as a tool for education and public outreach. It still continues in that role, serving as a powerful reminder of the persistent adaptability, curiosity, and resourcefulness of Wisconsin's astronomers, who helped create new fields of astrophysical research and whose work is represented today in space telescopes and modern great reflectors.

35

WISCONSIN WOMEN IN ASTRONOMY

Women's participation in astronomy has always been constrained by the social conventions and norms of wider society, many of which limited women's access to education itself.

Initially, women were considered by many male astronomers to be unsuited for the "physical hardships" of long nighttime hours at the telescope, not to mention academic life more generally. But norms and conventions shifted rapidly in the United States after the Civil War, and female students were becoming more common on college campuses by the time Washburn Observatory came along.

Despite critics who argued that rigorous study would adversely affect a woman's health or that women would distract men from their studies, female students across the nation grew in number, and the University of Wisconsin was among the more progressive in bringing women to campus. John Bascom was important in this as university president because he believed in equal access to higher education and academics for women.

Edward Holden was also very supportive of women who were interested in astronomy and employed several students not only for computation but for significant work with the observatory's new instruments. Among the most notable are Alice Sanborn and Alice Lamb, who published their work under their own names in the *Publications of the Washburn Observatory* between 1883 and 1886.

Computation was a highly skilled activity. Women's skills were equal to men's, Joel Stebbins remarked, except "usually the women are neater and just as accurate." However, before roughly the middle of the twentieth century, women who sought astronomical careers beyond computing were, with rare exceptions, limited to the handful of positions at small observatories at women's colleges, where they succeeded as practicing astronomers in defiance of biases. As happened with Lamb, social forces pressed many young potential astronomers, as Stebbins put it, to be "converted into matrimony," rather than pursuing careers in science.

In May 1963, these four women were graduate students in the UW Astronomy Department, making up an unprecedented one-fourth or so of the students pursuing a PhD in the field. Here they pose beside the Burnham telescope. From top: Susan Simkin, Laura Bautz, Bernadette Londak, and Natalie Satunas.
DEPARTMENT OF ASTRONOMY, UNIVERSITY OF WISCONSIN–MADISON

Betty Louise Webster, known as Betty Webster-Turtle after her marriage, primarily used Pine Bluff Observatory for her research work, though she is shown here at the eyepiece of the Washburn refractor.
DEPARTMENT OF ASTRONOMY, UNIVERSITY OF WISCONSIN–MADISON

Edith Flather, center, worked as a Washburn computer in the late 1950s and 1960s during the early days of electronic computers. (She would have been called a coder or computer programmer today.) On Flather's right is Wisconsin astronomer John Mathis, and on her left is visiting astronomer Rudolph Minkowski. In 1959, Flather and Morse Huffer co-authored a short history of Washburn Observatory, published by the Wisconsin Academy of Sciences, Arts, and Letters. DEPARTMENT OF ASTRONOMY, UNIVERSITY OF WISCONSIN–MADISON

After Holden's tenure, there seem to have been very few women employed at Washburn Observatory outside of the traditional roles of clerical help and computation. Before the 1960s, it was rare for the University of Wisconsin to award graduate degrees in astronomy at all. However, of the three graduate degrees awarded in astronomy in the first half of the twentieth century, one was a master of arts awarded in 1932 to Alletta Esther Beddoe.

While little remains about her in the historical record, Beddoe's hometown is listed in the university's 1932 commencement booklet as Springdale, Arkansas, and she received her bachelor's degree in 1930 from Carleton College, in Minnesota. Stebbins writes that Beddoe was working at Washburn during the summer of 1930, and he recommended her for an out-of-state tuition remission scholarship. Admissions forms from both 1930 and 1931 show her enrolled in a master's program, but, though the booklet lists her degree as astronomy, it does not mention her thesis title, and the nature of her work at Washburn remains obscure.

It wasn't until the fall of 1966 that Washburn Observatory employed its first female PhD astronomer on staff. Betty Louise Webster (1941–1990) came to Wisconsin from Australia for her postdoctoral research, specifically because it was the best place to advance her research in photoelectric photometry of planetary nebulae, a specialty of the university's Donald Osterbrock. An article in the *Capital Times* in 1967 shows how far views about women in observational astronomy had come: "Astronomy presents no special handicaps for women, least of all observations with the big telescopes, as Miss Webster sees things."

Women began to account for a significant portion of advanced researchers in astronomy in the early 1960s, and Wisconsin was part of that trend—about one in four graduate students were women. The first woman to earn a doctorate in astronomy at the University of Wisconsin was Natalie Satunas in 1964. Next, in 1967, came Susan Simkin (1940–2021), who became the first woman tenured in the Astronomy Department at Michigan State University, and Laura "Pat" Bautz (1940–2014), who spent much of her career at the National Science Foundation. There, Bautz eventually held the position of director of the Division of Astronomical Science, and she also served Wisconsin as a member of the Astronomy Department's Board of Visitors.

Faculty appointments of women in astronomy began to pick up beginning with astronomer Linda Sparke, a specialist in the dynamics of galaxies, who became a University of Wisconsin professor in 1989. Three decades later, roughly half of all PhDs awarded in astronomy at UW–Madison and elsewhere go to women. Their potential was overlooked for far too long.

36

WISCONSIN ASTRONOMY GOES TO SPACE

ARTHUR CODE

As Joel Stebbins developed photoelectric photometry and Albert Whitford introduced modern electronics to astronomy, Arthur Dodd Code (1923–2009) took Wisconsin astronomy into space.

Like his predecessors, Code was an innovator of the first magnitude. Originally from Brooklyn, New York, where as a child he became interested in astronomy, Code came to Wisconsin in 1946 as a graduate student at the University of Chicago's Yerkes Observatory in Williams Bay, Wisconsin. He was working on his PhD in astronomy when he made his first trip to Madison to attend the American Astronomical Society meeting in September 1946. By 1951, Code had joined the small staff of Washburn Observatory, alongside Director Albert Whitford and second astronomer Morse Huffer.

Code, an enthusiastic instrument builder with a solid background in electronics from his service in the navy during World War II, was in his element at Washburn. While photomultiplier tubes and chart recorders were quickly replacing the earlier photometric instruments that Stebbins and Whitford had developed, Code recalled being introduced to the fading era of photoelectric photometry as an observing assistant in the darkened observatory, "watching the dancing beam of light on the galvanometer scale."

Code's first stint at Washburn Observatory lasted until 1956, during which time he and Wisconsin graduate student Ted Houck worked together on a multifaceted observing program in South Africa, studying the structure of the Milky Way in collaboration with Yerkes astronomer William W. Morgan. Code and Whitford also worked together on the development of narrow-bandwidth photometric scanning techniques, which not only represented an advance in instrumentation but also were key to Code's critique that the controversial Stebbins-Whitford Effect was actually an instrumental effect, not a natural phenomenon.

By 1956, Code had been lured to a faculty position at the California Institute of Technology, where he had access to the 200-inch Mount Palomar telescope, then the largest in the world. But new technologies were rapidly developing, heralding the possibility of space exploration. Code recognized that it soon would be possible to put astronomical telescopes and instruments in space, and he knew what he wanted to do: measure the ultraviolet energy of hot, massive stars, which is undetectable by terrestrial instruments.

Arthur Code, around the time of his appointment as the sixth director of Washburn Observatory.
DEPARTMENT OF ASTRONOMY, UNIVERSITY OF WISCONSIN-MADISON

In late 1957, Code was chosen as Whitford's eventual successor, upon Whitford's impending move to Lick Observatory the following summer. Code would become the sixth director of Washburn Observatory and the last astronomer to be hired specifically for this position. Stebbins, by then a fellow at Lick Observatory, expressed approval and relief over Code's move back to Washburn, as it "eases our consciences here on taking Albert from Wisconsin."

Code's decision to return to Wisconsin was motivated by his plans to be a pioneer in the new field of space astronomy, an ambition that remained unattainable, in his view, in the land of giant ground-based telescopes. To attract Code to Washburn, the university promised him support for the new directions in astronomy he had in mind. Whether good fortune or brilliant foresight, that hiring decision proved to be a master stroke. In October 1957, the Soviet Union launched the world's first artificial satellite, commonly called Sputnik. That stunning technological triumph shocked the United States and started the space race, a twenty-year period in which the Americans and the Soviets would strive to outcompete each other in space technology and exploration. In July 1958, nine months after Sputnik's launch, Code assumed the Washburn directorship, Pine Bluff Observatory opened, and NASA was founded. Before long, Washburn's astronomical photometers would be heading for space.

Code was a member of the Space Science Board of the National Academy of Sciences, and in mid-1958, he was one of the first to propose specific plans for a scientific satellite. Weighing about one hundred pounds, Code's satellite would carry a small telescope collecting light to feed a Wisconsin photoelectric photometer, which would measure not visible light like its predecessors but the ultraviolet light of stars. NASA approved Code's plan, complete with funding, contracts, and deadlines, which meant rapid growth for the small Wisconsin astronomical community. They suddenly needed full-time engineers, computer programmers, project managers, and astronomers.

Today, satellites and space operations exist all around us as well-developed and familiar technologies, and spacecraft have visited every planet in our solar system. But in the late

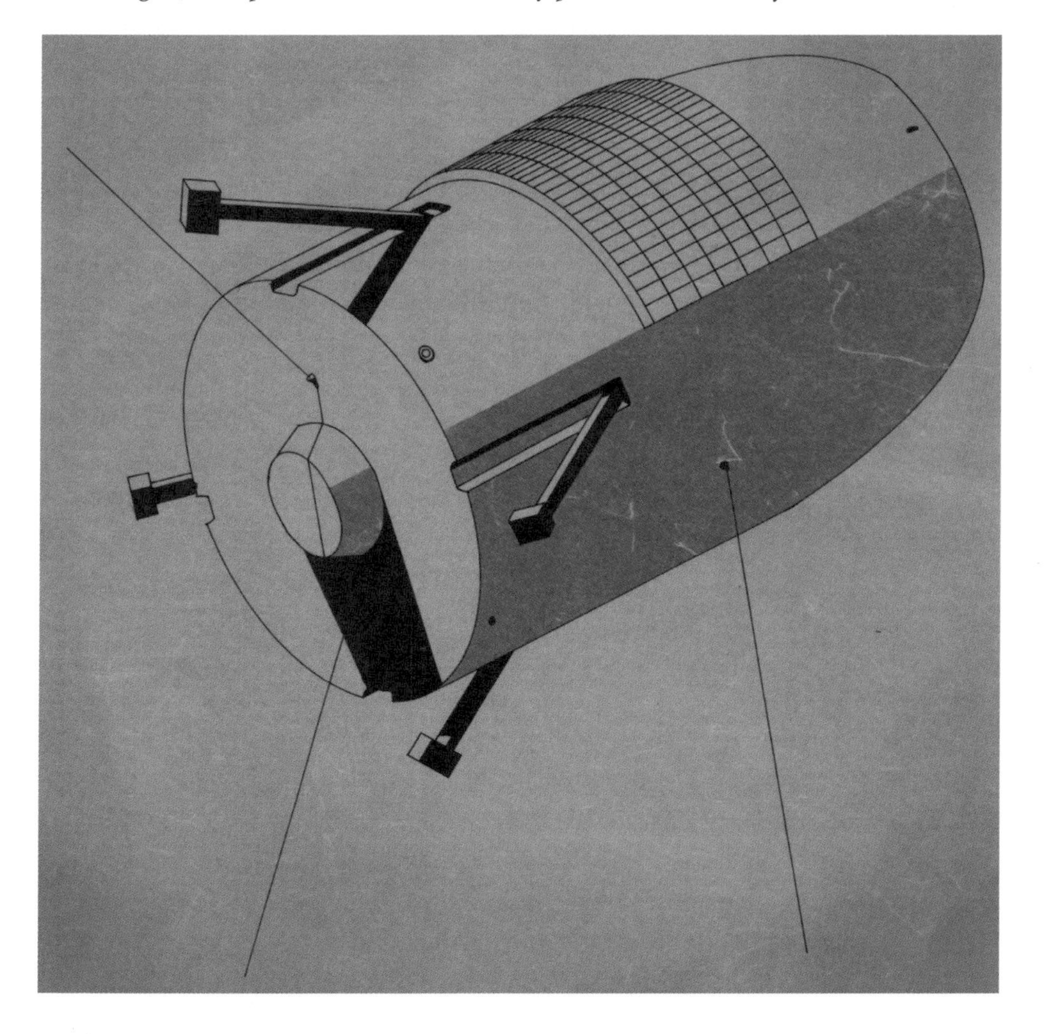

Code's concept for the one-hundred-pound satellite, shown in this sketch, consisted of a single small reflecting telescope that would collect ultraviolet light for an attached photoelectric photometer.
DEPARTMENT OF ASTRONOMY, UNIVERSITY OF WISCONSIN–MADISON

1950s, astronomical observing in space was a completely new enterprise that had never been done before, and the first space scientists were taking on daunting risks and challenges. Unlike any astronomer before him, Code proposed observing with a telescope that could neither be directly pointed nor adjusted, and with optics and instruments that had to survive a rocket launch, be calibrated and operated by some combination of remote and automated control, and be able to transmit their observations to the ground by radio. And it all had to happen in the then unknown but undoubtedly hostile environment of Earth's orbit. Though NASA eventually would provide solutions for many of the basic needs of space scientists, such as pointing a space telescope at astronomical targets, Code and his small team started with only their wits and the laws of physics. For example, in one of their early homemade experiments to understand pointing and guidance systems, they suspended a set of small gyroscopes from a long wire descending through a six-story stairwell in Sterling Hall. Starting small and understanding the basics would characterize Wisconsin's "space cadets" (as they sometimes called themselves, never ones to take themselves too seriously) while they proceeded to master new skills, such as building a vacuum-deposition chamber for making ultraviolet filters, designing flying photometers for balloons and rockets, and operating robotic telescopes.

To handle the rapid expansion, Code created the Space Astronomy Laboratory, colloquially known as SAL or simply the Lab, as a branch of the Astronomy Department, primarily to work on NASA-funded projects. Under Code's leadership, and with the help of other Wisconsin astronomers—notably Code's long-time colleague and collaborator Robert C. Bless (1927–2015)—SAL produced a steady series of successful astronomical instruments. Some flew on NASA's X-15 research rocket planes, many launched on suborbital rocket flights, some became major orbiting projects, and one astronomical instrument flew on two space shuttle missions.

Code's work in space astronomy would dominate his career, including his influential work in the planning of the space observatory that would become the Hubble Space Telescope and his appointment as founding director of the Space Telescope Science Institute. Like all of his predecessors as director of Washburn Observatory, Code was active and prominent in the astronomical community, being elected to the National Academy of Sciences (in 1971), and serving as president of the American Astronomical Society (from 1982 to 1984). Unlike any of his predecessors, Code was awarded in 1992 with the NASA Distinguished Service Medal.

37

ORBITING ASTRONOMICAL OBERVATORIES

Arthur Code's original concept for a space telescope was very simple: a single, small telescope, with an eight- or ten-inch mirror, collecting light for a photomultiplier detector tube. It would be powered by a few solar panels and send its measurements back to Earth by radio. The entire gadget would weigh about one hundred pounds and be about three feet long. NASA began funding Wisconsin's early development of such a flying photometer, but the first one to leave the ground actually went aloft on a helium balloon across Madison's Lake Mendota. Later experiments flew on NASA's X-15 rocket planes and on suborbital rocket flights as tests that not only produced scientific results but also paved the way for true orbiting instruments.

By the very early 1960s, the US space program had grown more confident and ambitious. NASA recognized that astronomical satellites, no matter their purpose, would all have certain needs in common, such as electric power, thermal management, communications, and especially the ability to point at and accurately track targets in the sky. These challenging problems were solved through the design of a common spacecraft that could be outfitted with various kinds of astronomical instruments. That common platform was called the Orbiting Astronomical Observatory, or OAO. The OAO satellite series paved the way for a far more sophisticated orbital photometer than Code's original one-hundred-pound concept.

With funding from NASA, astronomers at Wisconsin's Space Astronomy Laboratory got to work designing an instrument package for the first OAO satellite. With Code and colleague Robert Bless directing the project, and astronomers Ted Houck and John McNall leading the technical development, SAL designed a space-worthy package of five telescopes, each with its own photoelectric detector, and two scanning spectrophotometers, which were descendants of the ones developed by Code and Whitford in the early 1950s.

Arthur Code, Ted Houck, John McNall, and Robert Bless (pictured here from left to right) were the principal members and leaders of the SAL team that built the Wisconsin Experiment Package (WEP) for NASA's OAO satellites. They are posing in SAL's optical lab with one of the WEP stellar photometers, which comprises an eight-inch off-axis reflecting telescope and a photomultiplier tube detector, the housing for which is just above the upper circuit board, which Code is touching. In the background at the right, behind Bless, is the vacuum deposition chamber used to create the custom ultraviolet filters needed for the WEP instruments. DEPARTMENT OF ASTRONOMY, UNIVERSITY OF WISCONSIN–MADISON

Four of the telescopes, designed to study stars, were about the size of the single one that would have flown in Code's one-hundred-pound satellite. The fifth, huge by comparison, was a 16-inch nebular photometric telescope—nearly the same aperture as the venerable Washburn telescope—and was designed to study diffuse objects such as clouds of dust and gas. Each of the telescopes was equipped with a filter wheel, allowing the photometer to measure the light intensity in various sections of the visible and ultraviolet spectrum.

This versatile instrument package, firmly rooted in Wisconsin astronomical photometry, was known as the Wisconsin Experiment Package, or WEP. While it was designed in Wisconsin, WEP was mostly built by contractor Cook Electric in Chicago. Code and Bless chose Cook Electric over larger but more distant aerospace firms in part because of the proximity (they could drive to meetings), and also because their project, small by NASA and military standards, might not get adequate attention from the larger aerospace corporations.

The OAO satellites were designed with a central, hollow cylinder down the middle of the spacecraft that could accommodate two instrument packages. Code claimed to be the one who pointed out to NASA designers that because there is no up or down in orbit, the instruments could fit back to back inside the cylindrical chamber and look out in opposite directions, taking turns pointing at targets.

The first OAO satellite, known as OAO-1, was launched on April 8, 1966, carrying a WEP looking out one end and a package of X-ray detectors supplied by Massachusetts Institute of Technology at the opposite end. But these instruments never had the chance to be activated, because within hours of launch, the spacecraft's electrical system suffered a catastrophic failure and the entire satellite was destroyed.

NASA decided to try again, so it ordered a second WEP from Wisconsin's Space Astronomy Laboratory and Cook Electric. OAO-2 was launched on December 7, 1968, carrying the second WEP. Its partner this time was an ambitious instrument package from Smithsonian Astrophysical Observatory called Celescope, which was designed to produce ultraviolet images of the sky but had not been ready in time for OAO-1. Unlike its predecessor, the OAO-2 satellite, occasionally known by the nickname Stargazer, successfully began operations in space and became the world's first space observatory.

Like a ground-based observatory, OAO-2 could be pointed anywhere in the sky to observe with various configurations of instruments, depending on the target object and the purposes

Scan the QR code* to watch *OAO's First Scope in Space: Wisconsin Experiment Package*, a 1964 educational video produced by NASA.

*Or visit go.wisc.edu/wepandoao

Cleanroom technicians at Grumman Aerospace Corporation (above) in Bethpage, New York, work on one of the OAO series spacecraft. At the top of the spacecraft is the circular opening into the cylindrical chamber where the scientific payloads, such as the Wisconsin Experiment Package, later would be mounted. DEPARTMENT OF ASTRONOMY, UNIVERSITY OF WISCONSIN–MADISON

The Wisconsin Experiment Package is shown here after delivery to Grumman Aerospace and before its integration into the OAO. On the right half of the instrument, two of the four stellar photometer telescopes can be seen with their electronics modules and detector housings. DEPARTMENT OF ASTRONOMY, UNIVERSITY OF WISCONSIN–MADISON

The completed Orbiting Astronomical Observatory, in a NASA cleanroom, is being prepared for integration with the Atlas-Centaur launch vehicle. The solar panels and upper aperture door, shown deployed here, would be stowed into launch configuration before being installed in the payload shroud, which covered and protected the instrument during launch. DEPARTMENT OF ASTRONOMY, UNIVERSITY OF WISCONSIN–MADISON

of the astronomers. The instruments were managed from NASA's control center, located at Goddard Space Flight Center near Washington, DC, where teams of astronomers, engineers, and others assembled sequences of timed commands that were uploaded to the satellite and executed automatically—a mode of operation that the Wisconsin astronomers had been practicing with their robotic telescope in Pine Bluff. But the ever-changing demands of orbital operations, especially in the first months after launch, kept team members constantly busy, especially Houck, McNall, and their assistants, as they had to rewrite the command sequences every time an unexpected event changed the timing.

Exploring the universe in the previously unreachable ultraviolet part of the spectrum was sure to lead to new results, and the effort paid off handsomely. Among many other results: confirmation of the theory that comets are composed largely of water and other volatile ices and are thus possible sources of water during and after planet formation;

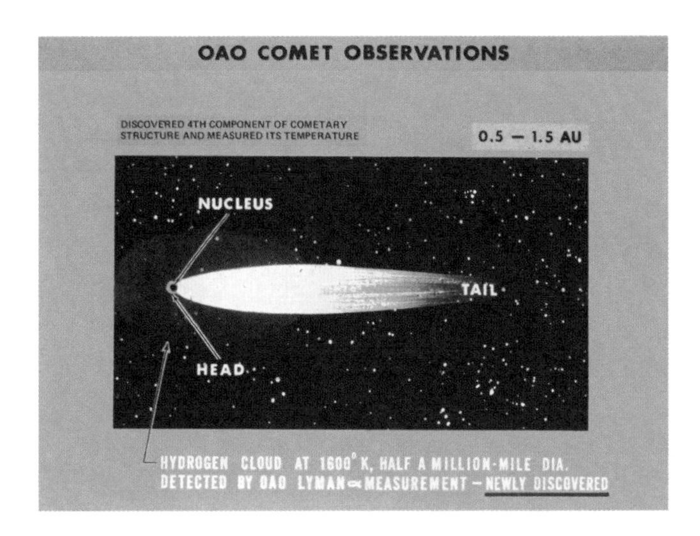

Astronomers used the Wisconsin Experiment Package on OAO-2 to observe comets, which enabled them to discover a vast cloud, or halo, around the comet's head. The cloud is visible only in UV light. Astronomers concluded that the cloud was caused by the breakdown of water molecules, showing for the first time that comet nuclei typically contain huge amounts of water. DEPARTMENT OF ASTRONOMY, UNIVERSITY OF WISCONSIN–MADISON

the discovery that radiation from expansion shells of stellar explosions can peak in the UV spectrum after they peak in the visible light spectrum, providing clues to the behavior of the clouds of matter that seed the interstellar medium with elements essential to planets and life; a new understanding of interstellar extinction of ultraviolet starlight (thus extending a field of study invented decades before by Joel Stebbins); and, of course, exploring one of the problems that first motivated UV astronomy, namely the nature and structure of the hot, massive stars that shape the evolution of galaxies.

As important as the science itself, OAO-2 demonstrated how valuable an orbiting observatory can be, ready to observe unpredictable events like new comets and stellar explosions, both of which came to pass on several occasions during the OAO-2 mission, including Comet Bennett in 1969 and Nova Serpentis in 1970.

Astronomers and spacecraft controllers also learned how to juggle scientific observations with spacecraft and orbital constraints—experience Code would take with him when he later served as the founding director of the Space Telescope Science Institute. OAO-2 operations were also made available beyond the science teams involved, so that guest observers could propose and carry out scientific observations using the space observatory—a custom continued today by the Hubble Space Telescope, among others. OAO-2's success put Wisconsin's astronomers and the Space Astronomy Laboratory on course to build other innovative space instruments that would become parts of the space shuttle and Hubble Space Telescope programs. OAO-2 operated successfully until it was shut down in February 1973, but the spacecraft and its instruments continue to orbit the Earth to the present day.

38

THE HEIR OF WASHBURN OBSERVATORY

SPACE ASTRONOMY LABORATORY

As Washburn astronomer Robert Bless liked to say, "The Lab was a remarkable place," but he didn't mean the Space Astronomy Laboratory's physical location. He meant "the Lab" as an operation, and especially as a team of talented people devoted to advancing astronomical research, especially through their expertise in developing instruments that could operate in the various challenging environments of spaceflight.

Bless and Arthur Code initially built the University of Wisconsin's Space Astronomy Laboratory around a small team of dedicated people, beginning in about 1958 with Ted Houck, who had recently received his PhD from the UW Astronomy Department. The Lab, as it was commonly known, was closely related to, but physically separate from, the Astronomy Department, permitting Code to better manage the new and more demanding projects that NASA was funding. Somehow Code and Bless, both of whom had worked in the old observatory before its move to Sterling Hall, captured everything that made Washburn Observatory remarkable—the innovative ethos, the persistence to work out the details of a project, the wisdom to keep things simple and build on local strengths, and the dedication to astronomical research that Washburn staff had steadily nurtured—and transplanted it all to a garage building on Madison's Park Street.

Beginning in September 1960, the Lab was located in the former Schmidt's automobile repair garage at 35 North Park Street, a few blocks south of the Astronomy Department offices in Sterling Hall (the Lab would remain there until 1973, when it moved to a campus space in the newly expanded Chamberlin Hall). The Lab grew quickly in the early 1960s, as it received NASA funding to develop the Wisconsin Experiment Package for the Orbiting Astronomical Observatory, as well as additional but related instruments for the X-15 rocket plane, suborbital rockets, and ground-based use.

The building that formerly housed Schmidt's auto garage was in poor repair when it became home to the Space Astronomy Lab in September 1960. Working conditions remained rough throughout the exciting years of NASA-funded work, until the Lab moved back to campus in 1973. WHI IMAGE ID 36110

In addition to astronomers Code, Bless, and Houck, the core staff in the Lab's early days included systems engineer and computer scientist John F. McNall, mechanical designer and machinist Richard Pfeiffer, optics specialists Dan Schroeder and Tim Fairchild, computer programmer Curt Heacox, and administrator and coordinator Kathy Brensike Stittleburg (who would later manage the Lab as associate director). Electronics engineer Donald Michalski and computer specialists Marilyn Meade and Tom Jones also joined the Lab during this era, along with many other highly skilled and dedicated researchers, technicians, and student assistants who often were entrusted with critical tasks in creating instruments and analyzing the resulting data.

In the late 1970s, the Lab expanded again. Code and Bless each submitted independent proposals to NASA for major astronomical instrument projects, and they thought they would be lucky if either proposal was chosen. To their surprise, NASA funded both. One proposal was for an instrument designed as part of the space telescope—now known as the Hubble Space Telescope—which was then in planning. The other proposal was for an instrument to be flown as part of astronomical space shuttle missions. The work of either project—much less both—would have fully occupied the Lab's capabilities, but Code had a philosophy he liked to apply to such situations: "When you come to a fork in the road, go both ways."

They decided that Bless would manage the space telescope project, developing an instrument called the High Speed Photometer (HSP); and Code would manage the other project, developing the Wisconsin Ultraviolet Photo-Polarimeter Experiment (WUPPE). The Lab lacked the space and staff to take on both major projects, so much of the design and construction of the High Speed Photometer was subcontracted to the university's well-equipped Space Science and Engineering Center, which specialized in instrumentation for a wide range of space missions from weather satellites to interplanetary probes. Meanwhile, WUPPE was designed and constructed almost entirely at the Lab, which by then had left the garage building and relocated to Chamberlin Hall.

Throughout this period, the Lab staff continued to work on suborbital rocket projects, which led to an important advance in general space instrumentation technology: the Star Tracker 5000. Lab astronomer Jeffrey Percival and engineer Kurt Jaehnig invented the new

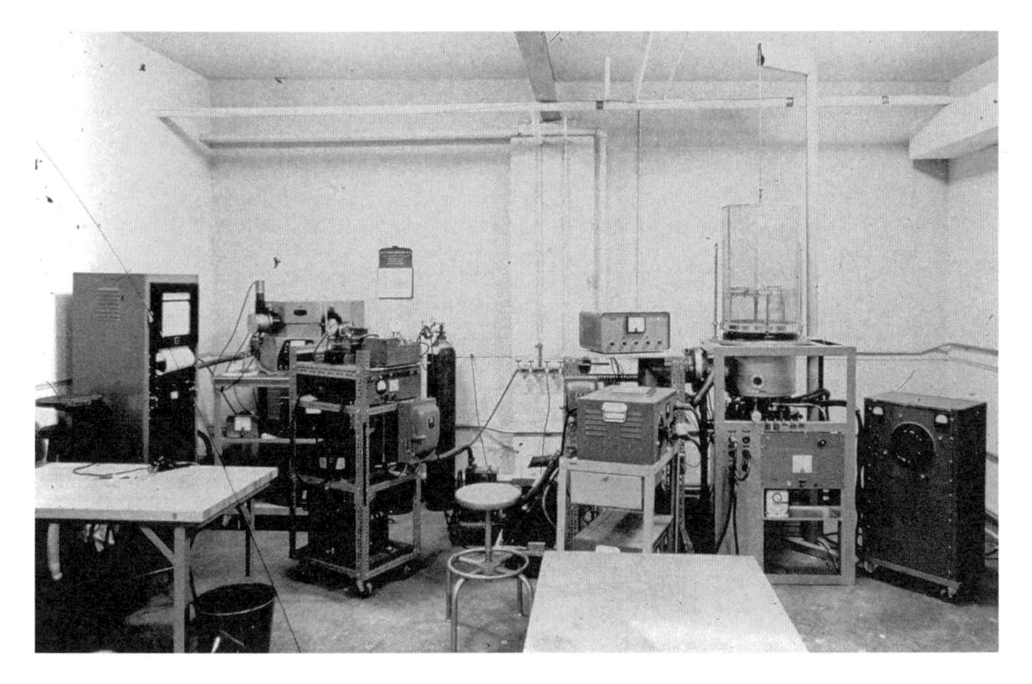

One of the many tasks of the SAL optics laboratory in the North Park Street days was making and testing the unique optical filters needed for instrumentation for the OAO and other projects. Near the far right is the cylindrical glass vacuum chamber, suspended from above, within which multiple layers of customized coatings were deposited on glass to make the filters. DEPARTMENT OF ASTRONOMY, UNIVERSITY OF WISCONSIN–MADISON

A few members of the High Speed Photometer project, in or around 1990. From left, astronomer and principal investigator Robert Bless, astronomers Matt Nelson and Jeffrey Percival, and engineer Andrea Tuffli. DEPARTMENT OF ASTRONOMY, UNIVERSITY OF WISCONSIN–MADISON

device to overcome the weaknesses of typical star trackers. The Star Tracker 5000 was designed to accompany an astronomical instrument on a suborbital rocket flight from launch to landing. It could recognize star patterns, calculate offsets and rotations, and send an image of the target field to the ground-based astronomer—more quickly, more accurately, and at less cost than the alternative devices that NASA was procuring from major aerospace contractors. The Star Tracker 5000 became widely used by NASA on rocket experiments and other flying scientific instruments, such as balloon-borne telescopes, that needed fine pointing.

Around the turn of the century, astronomical instrument development at the Lab began to turn away from space-based instruments, toward new, large, ground-based telescopes—the 3.5-meter WIYN telescope at Kitt Peak National Observatory in Arizona, and the ten-meter class SALT telescope in South Africa's Northern Cape province. To support this shift in work, the Lab was reorganized in 2013 to create Washburn Astronomical Laboratories, continuing the long Wisconsin tradition of developing highly advanced instrumentation for ultraviolet, optical, and, now, infrared astronomy. In effect, the original Space Astronomy Laboratory became, for a time, a subdivision of the new Washburn Labs.

Infrared astronomy is a good example of a Lab project that wasn't about hardware development. After the 2003 launch of the Spitzer Space Telescope, NASA's orbiting infrared observatory, Wisconsin astronomer Edward Churchwell led a major program funded by NASA and supported by Lab specialists to create a large-scale and highly detailed survey along the plane of our Milky Way Galaxy. Called GLIMPSE (for Galactic Legacy Infrared Mid-Plane Surveys Extraordinaire), the result was a three-dimensional exploration that penetrated to the far side of our galaxy, thus more than fulfilling the quest undertaken by

Joel Stebbins and Albert Whitford nearly eighty years before to identify and photometrically map the central galactic regions. The GLIMPSE data, images, and results are freely available to scientists and the public.

In October 2021, the Space Astronomy Laboratory was permanently shut down. While some of the work that was begun by the Lab will continue under Washburn Astronomical Laboratories, its closing marked the end of a highly consequential epoch in the long and ongoing tradition of astronomical instrumentation and science that grew out of Washburn Observatory.

An image from the GLIMPSE project reveals "bubbles" in the interstellar medium resulting from the intense radiation produced by very hot stars. NASA/GLIMPSE

39

FLYING TO THE EDGE OF SPACE

X-15 SPACE INSTRUMENTS

An orbiting satellite like the OAO is not the only way to get astronomical instruments into space. NASA's X-15 rocket-powered space-plane, made by aerospace contractor North American Aviation, was capable of flying at extremely high altitudes and was originally intended for engineering research. But in 1962, at the urging of Arthur Code and others, NASA expanded the research scope to include scientific experiments and modified the original X-15 design to accommodate a bay just behind the pilot compartment that could be opened for scientific instruments.

The first astronomical instruments to fly aboard the X-15 came from Code's Space Astronomy Laboratory, with the goal of testing the feasibility of techniques in ultraviolet astronomy. Unlike suborbital rockets of the early days, the X-15 could host a pointing platform for selecting specific stars, just as an astronomer can point a telescope at a selected target. The available observing time was limited to a few minutes, and the size and complexity of the instruments were tightly constrained, but it was a chance to put designs to the test—and, unlike a satellite, testing on the X-15 came with the huge advantage of getting the instrument back after the flight to check how it fared and recalibrate it in the lab.

In the early 1960s, Code and fellow astronomer Robert Bless were developing their instrument package for the first OAO satellite. Eager to gain experience designing flying astronomical instruments while simultaneously getting early data in preparation for OAO, Code and Bless received NASA funding to build and fly two astronomical instruments on X-15 flights. The first instrument was yet another variation on Wisconsin's astronomical strength in photoelectric photometry: a two-channel photometer that was sensitive in both visual wavelengths and the ultraviolet. The second was a scanning ultraviolet spectrophotometer, another version of the technique pioneered in the 1950s by Albert Whitford and Code.

NASA's X-15 suborbital spacecraft is seen in a NASA photograph with one of its pilots, Neil Armstrong, before his work in the Gemini and Apollo programs. Just behind the cockpit, the canopy of which is shown opened, was the instrument bay for scientific experiments. NASA

The two instruments flown by Wisconsin astronomers on the X-15, both made by the Space Astronomy Laboratory, were the UV spectrophotometer (on the left) and the two-channel photometer (on the right). The two-channel photometer contained two detectors to simultaneously measure the sky's brightness in both the visual and ultraviolet parts of the spectrum. Each "channel" of the photometer used miniature vacuum tubes, visible on the circuit board, for amplification. Out of six flights of the two-channel instrument, one of which ended in an X-15 crash landing, two produced useful data. (Once again, astronomer Ted Houck's OAO coffee mug is included for scale.) DEPARTMENT OF ASTRONOMY, UNIVERSITY OF WISCONSIN–MADISON

The detector end of the SAL X-15 UV spectrophotometer shows the miniature detector, a UV-sensitive photomultiplier tube. The base and electrical connections of the detector are embedded in white "potting" material. On the left is the small motor, which turns the gear above it. The gear turns a threaded rod, which moves the detector across the spectrum produced by the diffraction grating (not visible here). DEPARTMENT OF ASTRONOMY, UNIVERSITY OF WISCONSIN–MADISON

The Wisconsin instruments were designed to operate automatically when switched on by the X-15 pilot, at which time a pointing system, designed for the job by Milwaukee's Astronautics Corporation of America, would orient the instrument so data could be recorded. In 1962 and 1963, the Wisconsin two-channel photometer flew on six X-15 missions, the most dramatic of which occurred on November 9, 1962, when aircraft malfunctions led to an emergency landing and the X-15 rolled over, causing severe injury to its pilot, John McKay. The Wisconsin ultraviolet spectrophotometer flew on August 12, 1966, near the end of the X-15 era, and produced stellar data and measurements of the ultraviolet dayglow of the sky, which was essential to understanding the sky background against which orbiting UV instruments would have to observe. Soon to be phased out of flight by NASA anyway, the X-15 lost its advantages as a scientific platform as suborbital rocket payloads gained the ability to point at particular targets.

The SAL UV spectrophotometer, pictured here, is mounted in the X-15 experiment bay on the pointing system made by the Astronautics Corporation of America in Milwaukee. The spectrophotometer aperture, center, is surrounded by four modified 35-mm cameras to record the target field at various UV wavelengths.
DEPARTMENT OF ASTRONOMY, UNIVERSITY OF WISCONSIN–MADISON

SPOTLIGHT

WISCONSIN'S ASTRONAUTS

According to NASA, an astronaut is someone trained to travel and work in space. The word comes from Greek and translates to "space sailor." Meanwhile, astronomy—roughly the "laws" or "arrangements" of the stars—is a field of science dedicated to the study of the universe beyond Earth.

While the two are only marginally connected, it can be difficult to think about the subjects of interest to astronomy without thinking about astronauts. And Wisconsin can boast of several.

James Lovell. NASA

Perhaps the most well-known Wisconsin astronaut is James (Jim) A. Lovell, who took four trips to space—including two to the Moon—as part of the Gemini 7, Gemini 12, Apollo 8, and Apollo 13 missions. Born in Cleveland, Ohio, in 1928, Lovell grew up in Milwaukee in the 1930s and spent two years as a student at the University of Wisconsin, studying engineering, as part of his training through the Navy.

In April 1970, as commander, Lovell helped safely navigate the ill-fated Apollo 13 command module *Odyssey* to a successful splash landing on Earth with the crucial assistance of the mission's lunar module *Aquarius* flying in a previously untested configuration. With command module pilot John (Jack) L. Swigert and lunar module pilot Fred W. Haise, Lovell holds the record of farthest distance traveled away from Earth in space, at 248,655 miles. Though first uttered by Swigert, Lovell repeated and is often credited with the phrase, "Houston, we've had a problem here."

Donald "Deke" Slayton was born in Sparta, Wisconsin, in 1924. In 1959, Slayton was chosen to pilot the Mercury-Atlas 7 mission, but because of a heart condition discovered shortly before the flight, he did not make a space flight until 1975, when he served as docking module pilot for the Apollo-Soyuz Test Project mission. That flight represented the first successful test of a universal docking system and the first joint meeting in space of American astronauts and Soviet cosmonauts. Slayton served a number of leadership roles at NASA, including coordinator of astronaut activities and manager of orbital flight tests, before retiring in 1982.

Donald "Deke" Slayton. NASA

F. Curtis Michel, born in 1934 in La Crosse, Wisconsin, served as a scientist-astronaut for NASA in 1965. Astronaut Daniel Brandenstein, born in 1943, hails from Watertown, Wisconsin, and graduated from UW–River Falls. And while not from Wisconsin, Brewster H. Shaw—who flew his first space shuttle mission in 1983—earned bachelor's and master's degrees from UW–Madison.

Mike C. Lee, from Viroqua, Wisconsin, became a NASA astronaut in 1984 and traveled 13 million miles across four space flights, including a Hubble Space Telescope maintenance mission. Leroy Chiao, from Milwaukee, Wisconsin, became an astronaut in 1990 and served as the first Asian American mission commander as part of the Expedition 10 mission from October 2004 to April 2005. Jeffrey N.

Leroy Chiao. NASA

Williams, born in Superior and raised in Winter, Wisconsin, became an astronaut in 1996 and has spent 534 days in space, including a record-breaking mission on the International Space Station as part of Expedition 47 and 48.

In addition, two UW–Madison astronomers have been members of NASA's astronaut corps: Robert A. Parker served as mission specialist in 1983 on the Spacelab 1 mission and in 1990 on Astro-1. And Kenneth Nordsieck, one of the Space Astronomy Laboratory astronomers who designed and built WUPPE, served as backup payload specialist for Astro-1 in 1990.

Laurel Clark. NASA

Finally, astronaut Laurel Clark was born in Iowa in 1961 but, according to her NASA biography, she considered Racine, Wisconsin, her hometown. Clark earned a bachelor's degree in zoology in 1983 and a medical degree in 1987, both from UW–Madison, before serving in the Navy as an undersea medical officer and naval flight surgeon.

On January 16, 2003, Clark left Earth as part of the crew of the science-and-research-focused Columbia mission. Over the course of sixteen days, Clark and her crewmates completed about eighty experiments. However, on February 1, 2003, sixteen minutes before its scheduled landing back on Earth, the shuttle disintegrated as it entered Earth's atmosphere. Everyone on board perished.

In a final email sent to her family from space, Clark wrote:

> *I have seen some incredible sights: lightning spreading over the Pacific, the Aurora Australis lighting up the entire visible horizon with the city glow of Australia below . . . rivers breaking through tall mountain passes, the scars of humanity . . . a crescent moon setting over the limb of our blue planet. . . . Whenever I do get to look out, it is glorious. Even the stars have a special brightness.*

40

FIVE BUSY MINUTES IN SPACE

SUBORBITAL ROCKET ASTRONOMY

Launching astronomical instruments into space is a difficult problem that can be solved multiple ways. The X-15 rocket plane was an early solution, and it had some advantages, including the fact that the stresses of takeoff and landing were relatively light—mild enough for a human pilot to tolerate.

Another solution is the suborbital rocket—also called a sounding rocket—which simply flies nearly straight up until it reaches the edge of outer space, where it arcs over and then falls back without going into an orbit around Earth, which would take far more energy than the typical suborbital rocket can deliver. Along the top of its arc, the payload experiences several minutes of freefall above most of Earth's atmosphere, where it's free to observe the universe in the ultraviolet and other parts of the electromagnetic spectrum. Ideally, after reentering the atmosphere, a parachute returns the instrument to the ground intact, which means the astronomer can examine how well the instrument functioned in flight, perform postflight calibration, and perhaps even fly the instrument more than once. But the accelerations and vibrations at launch and reentry are tremendous, presenting considerable challenges for the designers of delicate instrumentation.

Wisconsin's Space Astronomy Laboratory developed many suborbital rocket instruments over the years, which NASA originally funded as preparation for the science and technology required for Orbiting Astronomical Observatory (OAO) operations. But the cost effectiveness of "rocket science," compared to satellite instruments, is so great that rocket programs continued well after the OAO program ended.

The first generation of Wisconsin rocket instruments flew on the liquid-fueled and highly reliable Aerobee rocket, of which there were several variants over the years. Wisconsin astronomers flew instruments on Aerobees from 1962 (the same year their X-15 instruments

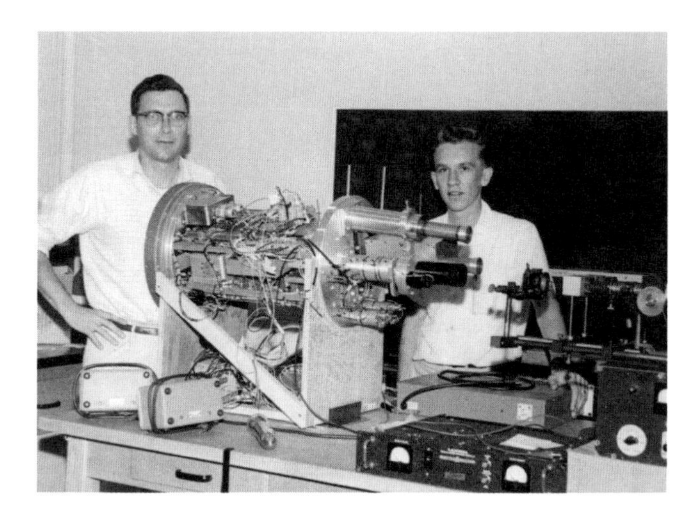

Robert Bless (left) and doctoral student George W. Collins II pose with the rocket instrument that flew on the first SAL Aerobee, mission 4.054, in October 1962. Launched from NASA's Wallops Flight Center, the instrument was created to test prototypes of modules that were being designed for the OAO project. UNIVERSITY OF WISCONSIN–MADISON ARCHIVES, MADISON, WISCONSIN

Astronomer Ted Houck, instrument maker Dick Pfeiffer, and astronomical systems engineer John McNall (from left to right), all of the Space Astronomy Laboratory, perform acoustic testing of an Aerobee instrument. DEPARTMENT OF ASTRONOMY, UNIVERSITY OF WISCONSIN–MADISON

began to fly) until 1979. Later generations of Wisconsin space instruments have launched on the solid-fueled Black Brant rocket and its variants. Both types of rockets have launched Wisconsin instruments into space from the Army's White Sands Proving Ground in New Mexico and NASA's Wallops Island Flight Facility off the coast of Virginia.

One of the Space Astronomy Lab's suborbital rocket instruments, the Wide-field Imaging Survey Polarimeter, was extremely successful, flying four suborbital missions—a lot for a suborbital astronomy payload. Known as WISP, the instrument was designed by Wisconsin astronomer Kenneth Nordsieck. WISP was a tour-de-force that showcased Wisconsin's astronomical strengths, bringing together ultraviolet photometric imaging with the measurement of the polarization of light from celestial objects, ranging from diffuse reflection nebulae

UW astronomer Robert Bless, on the right, oversees preparations for an Aerobee instrument launched on mission 4.055, in September 1964, from NASA's Wallops Flight Center. The purpose of the flight was to test prototypes of modules that were being designed for the OAO project. UNIVERSITY OF WISCONSIN–MADISON ARCHIVES, MADISON, WISCONSIN

Astronomers retrieve a suborbital rocket instrument that had successfully returned by parachute to the New Mexico desert after its launch from NASA's White Sands Test Facility during a mission around 1990. The person at far left is Wisconsin astronomer Ken Nordsieck. DEPARTMENT OF ASTRONOMY, UNIVERSITY OF WISCONSIN–MADISON

An Aerobee rocket is being transported to launch. DEPARTMENT OF ASTRONOMY, UNIVERSITY OF WISCONSIN–MADISON

to immense galaxies and a nearby comet. Polarization studies provide insight into the three-dimensional distributions of matter and the structure of particles such as interstellar dust.

NASA hastily planned one of WISP's four missions in April 1997 after Comet Hale-Bopp was discovered in 1995. The fact that WISP was ready to go off the shelf and could be flown on a suborbital rocket on relatively short notice (compared with much longer development times and greater costs for orbital instruments and launch vehicles) serves as testimony to one of the strengths of "rocket astronomy." WISP's observations of the ultraviolet sky background have helped astronomers distinguish the contributions of dust scattering from that of other background sources, such as numerous and distant quasars. WISP's most successful flight resulted in a new understanding of the Large Magellanic Cloud—a nearby satellite galaxy to our own Milky Way Galaxy. WISP also led to the development of the Lab's Star Tracker 5000, an important technology for pointing spaceborne instruments at targets.

Scan the QR code* to watch *Astronomy from the Edge of Space*, a video from the Space Astronomy Laboratory, produced and directed by Dave Potratz, about suborbital rocket science and SAL's WISP instrument.

*Or visit go.wisc.edu/edgeofspace

41

NASA'S HUBBLE SPACE TELESCOPE AND WISCONSIN'S HIGH SPEED PHOTOMETER

Historians point to three technologies as the tripod of modern astrophysics: photography, spectroscopy, and photometry. NASA's Hubble Space Telescope, launched in April 1990, carried that tripod into space in its ensemble of science instruments, which included cameras, spectroscopes, and photometers. Not surprisingly, Hubble's photometer package came from Wisconsin, where modern astronomical photometry was developed.

NASA's plans for a large space telescope began as early as 1968, and a decade later, with funding secured, the designing began in earnest. UW's Space Astronomy Laboratory proposed to NASA an instrument called the High Speed Photometer, or HSP, which would become one of the original instruments on the Hubble telescope at its launch. The HSP was designed to take full advantage of its position above Earth's atmosphere, which causes images to blur and shift in terrestrial telescopes. The HSP would be "high speed" because it could make very rapid photometric measurements, as fast as hundreds of thousands per second, which would allow astronomers to observe and understand rapidly changing and subtle phenomena, such as stellar explosions and disruptions by black holes or occultations of stars by planets and asteroids. And it might even have detected the first exoplanet transits, had its mission on the Hubble Space Telescope been completed. Hubble's promises of superior image quality and stability were supposed to be the keys to HSP's scientific success, but they proved instead to be its downfall.

Robert Bless and his team, which consisted of engineers and scientists at both SAL and UW's Space Science and Engineering Center, designed the High Speed Photometer to use then-state-of-the-art image dissector tubes (distant descendants of the detectors used by Joel Stebbins in the 1920s) that were capable of electronically sampling select parts of an image field. A clever arrangement of apertures and filters in front of the detector tubes made it possible to have a highly configurable instrument with no moving parts ("except

Starlight enters the Hubble Space Telescope from the left and is focused into the various science instruments inside the housing on the right. NASA

electrons," as Bless was fond of adding). HSP was the cheapest by far of Hubble's original science instruments, and the only one built entirely on a university campus with the help of many student employees.

HSP itself functioned perfectly after Hubble's launch—indeed by the end of its service on Hubble, it was the only science instrument not to have suffered a major subsystem failure. But the telescope itself (as distinguished from the individual science instruments) was famously a major failure.

The flawed primary mirror of the Hubble Space Telescope delivered fuzzy star images to all the science instruments, and the telescope pointing system had jitters that rendered even the blurred images unstable—problems that were devastating for HSP. Despite the telescope's severe shortcomings, Wisconsin astronomers still produced scientific results, including ultraviolet photometry of the fast pulsar in the Crab Nebula and detailed mapping of Saturn's rings using a stellar occultation. HSP also tracked the outburst of a nova.

Hubble Space Telescope's science instruments and other components were designed from the beginning to be removed in orbit by space shuttle astronauts during servicing missions. This capability was crucial to NASA's plan to fix the telescope's optical and pointing problems. During the first servicing mission, in December 1993, spacewalking astronauts installed a package of corrective optics to compensate for the flawed primary mirror; but to make space for the new optics, one of the primary science instruments had to go. NASA selected HSP

Astronomer Robert Bless helped found the Space Astronomy Laboratory. After growing up in Florida, Bless received his PhD in astronomy from the University of Michigan, then came to Wisconsin as a post-doctoral researcher in the fall of 1958, shortly after Arthur Code started as director. Bless first worked at PBO on a project to calibrate the absolute energy sensitivity of photometric detectors. By 1960, he had joined SAL and had been promoted to professor. Bless worked on OAO and suborbital rocket instruments, was the principal investigator of the High Speed Photometer, and was active in ground-based astronomy, helping create the twin eight-meter telescopes of Hawaii's Gemini Observatory. Unfailingly soft-spoken and modest, Bless was known among his friends as a Renaissance man—a popular teacher, textbook author, patron of the arts, wine connoisseur, historian, and good father. DEPARTMENT OF ASTRONOMY, UNIVERSITY OF WISCONSIN–MADISON

HSP Principal Investigator Robert Bless (left) and HSP Project Manager Evan Richards, of the UW Space Science and Engineering Center, pose in front of an engineering model of the HSP.

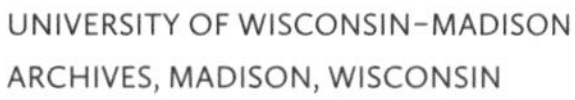

UNIVERSITY OF WISCONSIN–MADISON ARCHIVES, MADISON, WISCONSIN

HSP project engineer Andrea Tuffli in a cleanroom with HSP during testing and calibration at NASA's Goddard Space Flight Center. DEPARTMENT OF ASTRONOMY, UNIVERSITY OF WISCONSIN–MADISON

as the sacrificial hardware, thus terminating the potential for a great deal of fascinating science. This ended the scientific career of HSP. In a sense, it also marked the end of the era of astronomical photoelectric photometry, which stretched back more than a century, because astronomical instrument designers were quickly adopting the rapidly advancing technology of solid-state detectors in place of devices that use the photoelectric effect.

The space shuttle returned HSP to Earth, the first of the original science instruments to return from orbit, and NASA used it for a few years for engineering tests. In 2012, former members of HSP's original development team in Wisconsin, led by Space Science and Engineering Center engineer Fred Best, helped bring the retired HSP back to Madison, where it was placed on public display at the University of Wisconsin's Space Place (see "Welcoming the Public: Sharing Astronomy with the People").

42

WISCONSIN ULTRAVIOLET PHOTO-POLARIMETRY EXPERIMENT

Wisconsin astronomers have a long tradition of pushing available technologies to advance scientific research: James Watson's Vulcan telescope, even though unsuccessful, and George Comstock's Loewy device, which was quite successful, were both unconventional approaches to scientific problems that used specialized instruments. Joel Stebbins first developed his photoelectric photometry for the study of variable stars but found ways to apply it to new problems, such as the colors of stars and measures of solar activity.

The Space Age offered further avenues of research, in part because of increased access to the ultraviolet light emitted by celestial objects. Polarization is an important physical characteristic of light, generally caused by scattering of light by matter or by magnetic fields through which the light passes. In the mid-1970s, Wisconsin astronomers and engineers—led by Arthur Code and Kenneth Nordsieck—designed an instrument to measure polarization of ultraviolet light at a range of wavelengths. Comparing polarization in different wavelength ranges can give information about the otherwise invisible structure and mechanisms in stars, galaxies, and the interstellar medium. The measurement and study of the polarization of light waves is called polarimetry. Polarimetry can tell us about the structure of particles of interstellar dust, from which stars form, as well as the geometry of circumstellar disks, where planets form and where interacting binary stars exchange matter as they evolve.

The unique instrument for this job, designed to operate from the cargo bay of the space shuttle, was called the Wisconsin Ultraviolet Photo-Polarimeter Experiment—or WUPPE, universally pronounced "whoopee," for short. WUPPE was built almost entirely in the mechanical and electronic shops of the Space Astronomy Laboratory in Chamberlin Hall with some major structural components fabricated by the UW Physical Sciences

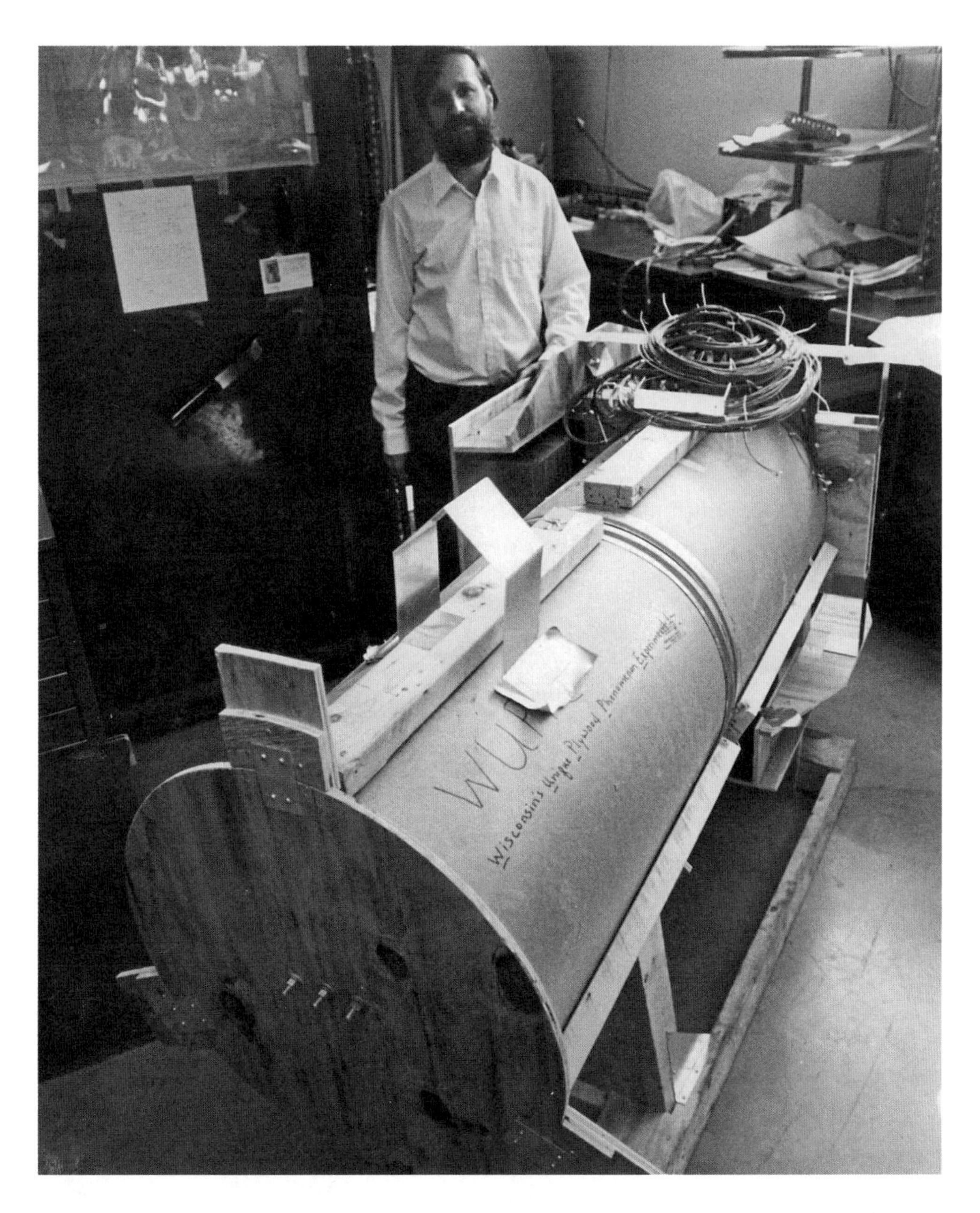

Astronomer Ken Nordsieck poses with a very early engineering mockup of WUPPE, built largely of wood and paper in one of the SAL shops. Before computer-assisted design software existed, a rough model like this helped engineers plan the three-dimensional arrangement of the instrument's components as constrained by the geometry required by the optical design. The hand-written label on the side reads "Wisconsin's Unique Plywood Phenomenon Experiment." UNIVERSITY OF WISCONSIN–MADISON ARCHIVES, MADISON, WISCONSIN

Laboratory. All the components were assembled and tested in a cleanroom across the hallway from the shops.

As with its sibling instrument, the High Speed Photometer, NASA selected and funded WUPPE to operate as a member of an ensemble of spaceborne instruments. WUPPE's ensemble was intended for a series of shuttle flights dedicated to astronomical research, as many as six, called the Astro missions, planned to begin in 1986. While the High Speed Photometer shared the Hubble Space Telescope's optical system with its companion instruments, the Astro instruments each had their own optical assemblies and instead shared the mounting and pointing systems. So WUPPE was a largely self-contained unit consisting of a 0.5-meter reflecting telescope feeding light to a photo-polarimeter—a distant descendent of the photometers that once had been used with the Washburn refractor.

Unlike the crewless Hubble Space Telescope, astronomers were included as Astro crew members to operate the astronomical instruments. NASA selected Ken Nordsieck, UW astronomer and WUPPE designer, to fly as one of those payload specialists.

After its second orbital mission, WUPPE returned to the Space Astronomy Lab and the Chamberlin Hall cleanroom where it was built, but was no longer subject to cleanroom protocols since it would never be launched again. Posing with their veteran space telescope in the background are Wisconsin astronomers leading the WUPPE team, from left, Ken Nordsieck, Arthur Code, and Chris Anderson. SAM GABELT

This striking photograph was taken by astronaut Sam Durrance during the first Astro mission, STS-35, in December 1990. It shows the Astro payload instruments fully deployed for operation (with an additional outline to distinguish WUPPE from the rest of the package). The view is over the payload bay through a window in the aft flight deck of the space shuttle orbiter *Columbia*. The right half of the image is dominated by the constellation Orion. The large instrument left of WUPPE is the Hopkins Ultraviolet Telescope. NASA

After delivery to NASA, WUPPE was integrated with the other instruments and pointing system and then mounted in the payload bay of the space shuttle orbiter, in preparation for the first Astro mission, scheduled for March 1986. But those plans were abandoned after the *Challenger* disaster in January of that year. No space shuttle flew again for nearly two years while NASA investigated the accident and extensively rescheduled planned missions. In the end, the number of Astro missions was reduced to two.

After nearly four years in storage, followed by many other launch delays, the Astro-1 mission finally flew in December 1990 on the space shuttle orbiter *Columbia*. Nordsieck was assigned to that mission as backup to one of the payload specialists. Despite a series of technical problems, Astro-1 accomplished history's first astronomical polarimetry in ultraviolet light. Astro-2, the second and last mission for WUPPE, launched in March 1995 on the orbiter *Endeavour*.

In February 2022, members of the Astro Restoration Project team completed the reintegration of WUPPE onto the instrument structure called the cruciform, which supported all the Astro instruments during preparation, launch, flight, and return. WUPPE was the first of the Astro instruments to be remounted in the restoration and integration work at the US Space and Rocket Center in Huntsville, Alabama. JAMES LATTIS

In addition to data on interstellar dust clouds in our own galaxy and in the Large Magellanic Cloud, WUPPE detected what is called bipolar flow, produced from the disks of interacting binary stars, which are highly evolved pairs of stars exchanging matter with each other. This was the first demonstration that such bipolar flows—known to happen at a vastly larger scale in galactic nuclei in exploding stars and in protostars—are also produced in the circumstellar disks of interacting binaries. Wisconsin's Pine Bluff Observatory played a role by complementing WUPPE's orbital ultraviolet observations with polarization data in visible light. Pine Bluff's 36-inch telescope, with an attached visual polarimeter built at the Space Astronomy Lab, gathered data on hundreds of targets, many the same ones observed by WUPPE, and many of those were nearly simultaneous observations, which is important in the case of variable stars, for example.

After the second and last Astro mission, WUPPE was returned to the Lab for postflight calibration and other work, after which the instrument went on public exhibit at UW Space Place. In 2021, WUPPE was lent to the US Space and Rocket Center in Huntsville, Alabama, where the Astro Restoration Project is working to reunify the companion Astro instruments and reintegrate their mounting, pointing system, and space shuttle pallet hardware on which they flew during the Astro missions. WUPPE and the unified Astro payload eventually will go on public display at the Smithsonian National Air and Space Museum in Washington, DC.

SPOTLIGHT

WELCOMING THE PUBLIC

SHARING ASTRONOMY WITH THE PEOPLE

Well before the Wisconsin Idea in 1904 declared that the borders of the university are the borders of the state, Wisconsin astronomers were already engaged in outreach to the public.

Legend has it that Governor Cadwallader Washburn wanted his new observatory to be accessible not only to university astronomers but to the people of Wisconsin. No documentation of that wish seems to survive, but Washburn lived to see public access become a fact.

From its beginning in 1879, the Washburn Observatory and its Clark 15.6-inch refractor attracted immense public interest. Barely a month after the telescope was installed—then the third largest in the country—both houses of the state legislature visited to look through the great equatorial. Though the first director, James Watson, came to Madison with a reputation for being stingy with observatory access (even for students), prominent visitors were occasionally welcomed to view celestial objects through the telescope.

After Watson's death in 1880, Washburn's second director, Edward Holden, wasted little time in expanding public access to the new observatory. On the first and third Wednesday evenings of each month, weather permitting, Washburn astronomers would welcome the public, though visitors would need to leave by midnight so the night's research could resume. Diverting from observational work for two evenings a month was highly unusual for research observatories of the era and testified to the university's intention to serve the residents of the entire state. Holden's twice-monthly visitor nights, which were announced with some solemnity in the newspapers, went into effect on April 20, 1881, creating a tradition still faithfully observed at Washburn Observatory today.

Observatory log books from 1881 forward often mention the visitor nights, noting crowds of one hundred or more people and sometimes even occasional damage and minor theft by visitors in the dark. Not all visitors followed the schedule. On a Thursday evening in October

1889, about thirty visitors attending the Good Templars convention in Madison, inspired perhaps by a waxing gibbous Moon, knocked at the observatory door. Assistant Sidney Townley, there to carry out his time service duties, let them in and showed them the large telescope, through which they observed the Moon. Under-supervised school children became a problem for Joel Stebbins in 1924, who wrote to the principal of nearby Randall Elementary School asking that teachers not bring their classes to visitor nights because they were showing up in unmanageably large numbers. Instead, Stebbins offered to arrange special observatory nights with individual teachers.

Particular celestial events produce large public demand for a glimpse through the telescope and often require deviating from the schedule. Favorable oppositions of Mars, Jupiter, or Saturn, for example, especially if mentioned in the media, can produce long lines. One can easily imagine how the close approach of Comet Halley in 1910, which was widely anticipated and frighteningly discussed in the media, produced a frenzy of interest that was hard for the astronomers to manage. Director George Comstock created a ticketing system to manage the large crowds and, recognizing an opportunity, got permission from the university to charge a small admission fee to defray the expenses of hosting visitors. As Stebbins later wrote:

> *On one of the days during this rather hectic astronomical period, [Comstock] was called to the telephone by an inquiring taxpayer. The question was: "Professor, what are you going to do with the money you are collecting for a view of Halley's comet?" Promptly came the response: "Madam, we are going to get a new tail for Halley's comet." The reply seemed to be entirely satisfactory.*

In the early days when the astronomical staff was small, visitors might be welcomed by Comstock, Stebbins, Albert Flint, or Morse Huffer. But by the early 1960s, the hosting of visitor

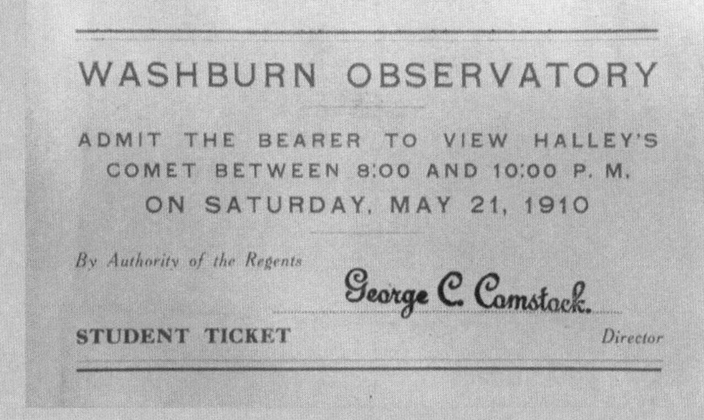
WASHBURN OBSERVATORY

ADMIT THE BEARER TO VIEW HALLEY'S COMET BETWEEN 8:00 AND 10:00 P. M. ON SATURDAY, MAY 21, 1910

By Authority of the Regents

George C. Comstock.

STUDENT TICKET *Director*

The return of Comet Halley in 1910 incited huge public interest worldwide, and astronomers everywhere responded. At Washburn Observatory, a ticketing system was instituted to control the number of visitors who wanted to see the comet. UNIVERSITY OF WISCONSIN-MADISON ARCHIVES, MADISON, WISCONSIN

nights increasingly became a task for the growing number of astronomy graduate students. More than one hundred forty years later, Holden's policy remains in place, expanded during the summer months to include weekly Wednesday evenings if the sky is clear.

Wisconsin's major contributions during the Space Age, combined with the great public curiosity about space astronomy, ushered in a new form of outreach that would leverage the enormous educational and inspirational potential of spacecraft, astronauts, and the exploration of the cosmos. In 1990, Arthur Code and fellow Space Astronomy Laboratory staff Robert Bless and Kathy Stittleburg created UW Space Place, an exhibit and education space where schoolchildren and the general public could explore the technology of working in space, see examples of spaceflight hardware, and learn about the science of space astronomy, often from the scientists themselves.

Space Place opened in July 1990, in a rented building on Madison's Park Street, a mile or so south of the Space Astronomy Lab's original location. By 1995, Stittleburg had acquired from NASA a full-scale engineering model of the OAO-2 satellite and installed it as an impressive

Washburn Observatory still offers visitor nights on the same schedule instituted in 1881, and with the same telescope. The modern schedule includes additional nights in the summer. JEFF MILLER, UNIVERSITY COMMUNICATIONS

At Family Science Night at UW Space Place, visitors explore astronomy through interactive learning experiences and creative opportunities, such as creating personal constellations and imagining an alien life form. UNIVERSITY COMMUNICATIONS

UW–Madison astronomer Blakesley Burkhart (right) speaks to a group in a Wisconsin state park for Universe in the Park. UNIVERSITY COMMUNICATIONS

exhibit. Soon afterward came WUPPE, the first of several veteran space instruments to go on public display. Each year, Space Place welcomed thousands of visitors for talks and workshops about space astronomy and for "star parties" celebrating lunar eclipses, meteor showers, Mercury transits, and other astronomical events.

In the summer of 2005, Space Place relocated a mile down the road to a larger building with a dedicated classroom, lecture hall, and rooftop observing deck. Space Place remains there today, offering exhibits and expert lectures, as well as programs about celestial events, star gazing, and amateur astronomy. Space Place's community education projects include a scale model solar system extending twenty-three miles from Madison to Mt. Horeb, and many other programs for schools, libraries, community centers, and parks.

In the mid-1990s, Wisconsin astronomers introduced yet another outreach program, called Universe in the Park. The idea is a simple one. At parks across Wisconsin, small teams of Wisconsin astronomers give a talk for visitors about an aspect of astronomical science or astronomical events, then set up a portable telescope for an evening "star party." This is

UW–Madison astronomer Paul Sell assists visitors with a portable telescope at Universe in the Park.
UNIVERSITY COMMUNICATIONS

perhaps astronomy education in its purest form, because the intrinsic beauty of the night sky leads naturally into inquiry and inspiration.

Universe in the Park started with WUPPE astronomers Karen and Jon Bjorkman, thanks to a modest outreach grant from NASA. Since then, the program has reached thousands of park visitors and campers, largely under the guidance of Wisconsin astronomer and Letters and Science dean Eric Wilcots. "The best environment in which to introduce the general public to astronomy is outside under dark skies," Wilcots has said. With an astronomer as a guide and a telescope as a tool, an evening under the stars can introduce us to a universe that scientific investigation reveals as more complex and fascinating every day.

See "Public Resources" to learn more.

43

THE ARC OF HISTORY BENDS HERE AND THERE

Observatories and universities, like other human organizations, necessarily evolve if they last very long and accomplish very much. The forces that drive this evolution are both internal, such as the ever-changing mix of people who collaborate to make an institution, and external, such as the changing character of a field or industry and the shifting priorities of funding sources. Such evolution is evident in the history of Washburn Observatory, starting with the "old" astronomy, to the rise of astrophysics, and through the development of space astronomy. The individual successes and failures, the discoveries and the inventions, and the progress toward a greater scientific understanding of the universe all form parts of a historical narrative from which we can learn.

The development of astronomical photoelectric photometry—encompassing not only the technical work of developing new instruments but also the scientific applications and innovations those instruments made possible—was the pivotal internal dynamic of Washburn's evolution. That development stemmed from a long-standing willingness to try something new—essential for any astronomical innovation—dating back to James Watson's day. When it all works, innovation produces scientific results that, in addition to contributing to our deeper understanding of nature, frequently raise new questions that lead to future innovations, forming a sort of virtuous circle. Then came the Space Age, a huge external force, which drove a new phase of Washburn's evolution by opening up new dimensions of scientific potential for astronomical photometry, while bringing new challenges, demands, and constraints.

The success of Wisconsin photometry under Joel Stebbins was widely admired and emulated. Yerkes, Lick, and Mount Wilson Observatories, among others, copied Wisconsin's instruments and worked with Wisconsin astronomers to make use of them. Other prominent astronomers, such as George Ellery Hale, Walter Baade, and Edwin Hubble, regularly sought the expertise of Stebbins and his team, funded their work, or encouraged their innovations. But Stebbins did not really develop a school of photoelectric photometry at Washburn Observatory. He thought the observatory faculty (which consisted only of himself, Morse Huffer, and later Albert Whitford) was too limited to offer proper graduate studies in astronomy. In fact, Stebbins and Whitford each supervised only two doctoral students in their entire Wisconsin careers, during which time Washburn was an observatory focused primarily on research. Only after the expansive growth of the university after World War II, the formation of the Astronomy Department in 1948, and the rapid increase in faculty under Arthur Code's directorship did Wisconsin's graduate program in astronomy begin to grow.

As the Astronomy Department expanded, Code saw to it that the lively tradition of instrument development, no longer based at Washburn Observatory, was cultivated at the Space Astronomy Laboratory. Following the examples of Stebbins and Whitford, the Lab's ethos valued simplicity in organization, instrumentation, and research programs. (Code considered organizational charts anathema, according to Robert Bless.) This translated into simple, reliable instruments, realistic timelines, and low budgets. The Lab, largely funded by NASA and relatively removed from academic cycles and concerns (aside from the lives of the always essential graduate students), pursued home-brewed innovation where necessary.

These characteristics greatly contributed to the long series of successes in suborbital rocket instruments, which helped advance the research programs of the Physics Department's Space Physics division, whose scientists were often working shoulder-to-shoulder with the astronomers. The virtues of simplicity and self-reliance also worked when scaled up to the much larger projects of OAO satellites, the High Speed Photometer, and WUPPE. Robert Bless was fond of pointing out that the High Speed Photometer was delivered to NASA on time and under budget, was built with student help (he liked to mention at NASA meetings that one of his student employees was a medievalist), and worked flawlessly, in considerable contrast to the other first-generation science instruments built for the Hubble Space Telescope.

As external funding patterns shifted, largely at NASA, no suite of internal virtues could sustain an operation like the Space Astronomy Laboratory for long, and some, such as the Lab's small size, could be distinct disadvantages when competing for new projects. The end of the space shuttle era reduced flight opportunities for some types of payloads. As ever-larger fractions of NASA's space science budget were devoted to huge, complex projects, such as the James Webb Space Telescope, funding for smaller space astronomy projects, such as

The WIYN 3.5-meter telescope, a modern reflecting telescope with a mosaic mirror, has been in operation since 1994. The primary mirror has actuators that actively control its optical shape for the best performance. A companion 0.9-meter WIYN telescope, also managed by Wisconsin astronomers, shares the site at Kitt Peak National Observatory, outside of Tucson, Arizona. Instruments from the Space Astronomy Lab and the Washburn Astronomical Lab are an important part of the astronomical capabilities there. NOIRLAB UNDER CREATIVE COMMONS

suborbital rocket instruments, was increasingly crowded out. Much of the funding that remained went to large aerospace companies instead of university laboratories.

At the same time, especially in the late 1990s and early 2000s, the optical research interests of Wisconsin astronomers shifted toward large, ground-based telescopes, as the university entered into consortium arrangements for such major observatories as WIYN (named after original consortium members Wisconsin, Indiana, Yale, and the National Optical Astronomy Observatory), with a 3.5-meter telescope, and SALT (the Southern African Large Telescope), with a ten-meter class telescope. In a sense, the reorientation to ground-based instrumentation was a return to the patterns of the Stebbins-Whitford era. As new fields of research emerged, Wisconsin astronomers suddenly had very different technological needs, which ultimately inspired evolution yet again when the Space Astronomy Lab became a subset

The Southern African Large Telescope (SALT) was built by an international consortium in which the University of Wisconsin–Madison is the largest nongovernmental partner. Located in South Africa's Northern Cape Province, SALT has been operational since 2011 and is the largest telescope in Earth's Southern Hemisphere. It has a ten-meter class primary mirror comprising a mosaic of 91 hexagonal mirrors, each of which is adjusted relative to its neighbors to act as a single large mirror. Wisconsin's Space Astronomy Lab and Washburn Astronomical Labs designed and constructed the primary instruments for astronomical research at SALT.
JEFF MILLER, UNIVERSITY COMMUNICATIONS

UW–Madison joined the global partnership to build SALT because of its expertise, tracing back to Joel Stebbins, in designing and building astronomical instruments. "We were the first place known for doing photoelectric photometry," says Eric Wilcots, UW–Madison astronomy professor and former SALT board member. Wilcots, pictured here with former graduate student Julie Davis, has performed research using the SALT telescope, which is purportedly so powerful it could detect the flicker of a candle on the moon.
JEFF MILLER, UNIVERSITY COMMUNICATIONS

of the newly formed Washburn Astronomical Laboratories in 2013. The new Washburn Lab carries on Washburn Observatory's original strengths in high-level astronomical instrumentation and its spirit of innovation—for example in state-of-the-art spectral instrumentation like the Robert Stobie Spectrograph for the SALT ten-meter telescope, as well as crucial components for the extreme precision NEID Spectrograph (from Penn State) for the WIYN 3.5-meter telescope.

When Stebbins brought photoelectric photometry to Wisconsin in 1921–1922, he very quickly transformed a classic nineteenth-century observatory, thoroughly based in the

Washburn Astronomical Labs staff prepare the NEID port adapter for installation on the WIYN 3.5-meter telescope in 2019. From left to right are opto-mechanical engineer Michael Smith; Marsha Wolf, principal investigator and director of WAL; and astronomical systems engineer Kurt Jaehnig. NEID is a high precision radial velocity spectrograph built by Penn State University for the WIYN Telescope to study exoplanets. WAL built the all-important port adapter, the interface between the WIYN telescope and NEID. The port adapter conducts essential tasks for exoplanet observations: It acquires the image of the target star, precisely positions that image onto a tiny optical fiber that sends the light to the spectrograph, actively maintains that positioning during the observation, and compensates for atmospheric dispersion caused by the Earth's atmosphere. After the observation is complete, the port adapter sends calibration light down the fiber in exactly the same way, enabling precise scientific measurements that reveal information about the exoplanet. NOIRLAB UNDER CREATIVE COMMONS

astronomy of precision and already halfway to obsolescence, into a force in modern astrophysics. At the same time, he launched a technological program, accelerated by his successors, that put Wisconsin astronomy on the frontiers of electronic instrumentation and space science during the twentieth and early twenty-first centuries. In a metaphor employed by novelist Thomas Pynchon, "gravity's rainbow" tracks the ascent, culmination, and descent of a suborbital rocket trajectory—a fine metaphor as well for the story of Washburn Observatory and the Space Astronomy Laboratory. The fact that such an arc must end is no reflection on the accomplishments and successes during flight.

The nature of astronomical science necessarily changes as our technical capabilities progress and our understanding of the universe grows. When Cadwallader Washburn gave an observatory to Wisconsin, he was, in part, repaying a debt to the place and the people who had made possible his great success. But he was also establishing a trust for the future, providing posterity with the means to achieve their own successes and adapt to new knowledge and capabilities. We, the incarnation of that future, inherit that debt as well, and we pay it by investing in our own future, wherever it will go, in our turn. From Observatory Hill, to Pine Bluff, to orbiting telescopes, perhaps someday to lunar observatories, and whatever comes after that, we will seek to explore and better understand our universe. It is up to us, to paraphrase Stebbins, to ensure that future nights will be both clear and dark on those new Observatory Hills.

The trajectory of a rocket launch produces a beautiful arc across the night sky. BEN SMEGELSKY, NASA

ACKNOWLEDGMENTS

This book could not have been created without the assistance and expertise of many people. Foremost we must thank the patient and proficient staff of the University of Wisconsin Archives for their many years of help and for sharing their deep knowledge of the collections in their care. Katie Nash, Cat Phan, Sarah Grimm, and David Null all surpassed the call of duty in our interest. We owe similar gratitude to the University of Wisconsin–Madison Memorial Library Department of Special Collections, the Wisconsin Historical Society Archives, and the Special Collections and Archives of the University of California, Santa Cruz, which holds the Lick Observatory Records. On the history of buildings and grounds at the University of Wisconsin, Daniel Einstein, historic and cultural resources manager in the University of Wisconsin Division of Facilities Planning and Management, has often and generously shared his expertise. This work has also made good use of NASA's Astrophysics Data System.

In years past, Catherine Scherb and Peter Susalla sorted, indexed, and preserved a large store of previously neglected historical materials relating to Washburn Observatory and Wisconsin astronomy more generally. Their efforts in preserving and bringing that material to light made it both accessible and ultimately suitable to transfer most of it to UW Archives. In addition, Peter's original research and writings underpin a number of the subjects treated in this book. Historians Jordan Marché II, David S. Liebl, and Dana Freiburger have been invariably helpful, generous, and collegial in our mutual interest and writings in the history of Wisconsin astronomy. Thanks are also due to Peter Sobol and Jay Gallagher for helpful discussions defining the ultimate scope of this history.

The members of the Antique Telescope Society have been a unique and invaluable resource and generous with their expertise in the technical details of vintage astronomical instruments. And, similarly, we have often benefited from the deep knowledge of local history shared by members of the Madison Astronomical Society, especially its *de facto* historian, John Rummel. Sam Gabelt, of the Space Astronomy Laboratory, provided important help with photographs from his collection. For their personal recollections of their families, we thank Bill Whitford (son of Albert Whitford) and Janet Dodge (granddaughter of Joel Stebbins).

For their assistance with archival work and careful reading of drafts, we thank Tony Lattis and Jacqui Beran. Further, we have benefited from Tony's historical work and publication on the Wisconsin collection of astronomical photodetectors, now held by the Smithsonian National Air and Space Museum, as well as the efforts and interest of David DeVorkin, of the same institution, who created the detector collection and has assisted at many points in the preservation and understanding of Wisconsin astronomy.

For background, thoughtful conversation, and links to primary research materials, we thank Sissel Schroeder, UW–Madison professor of anthropology, and Omar Poler, Indigenous education coordinator in the UW–Madison Office of the Provost. We also recognize with gratitude the indulgence of UW–Madison Astronomy Department chairs Robert Mathieu, Ellen Zweibel, Sebastian Heinz, and Richard Townsend, as well as the staff and leadership of University Communications, especially senior photographer Jeff Miller and science writer Elise Mahon. In the indulgence department, thanks are also due to Jennifer Lattis, James Tyrrell, and Kay Kriewald (who keeps UW Space Place running). Finally, we acknowledge with thanks the editorial staff of the Wisconsin Historical Society Press for their patience, assistance, and transformative improvements in the content and form of this book. Any errors or deficiencies are, of course, the responsibility of the authors.

WHI IMAGE ID 24999

PUBLIC RESOURCES

Learn more about events, courses, and other opportunities for K-12 students and the general public:

Bell Burnell Observatory (formerly the Washburn Student Observatory)
www.btci.org/bell-burnell-observatory

Washburn Observatory Visitor Nights
http://go.wisc.edu/washobs

UW Space Place
www.spaceplace.wisc.edu

Universe in the Park
www.astro.wisc.edu/outreach/universe-in-the-park

UNIVERSITY OF WISCONSIN–MADISON
ARCHIVES, MADISON, WISCONSIN

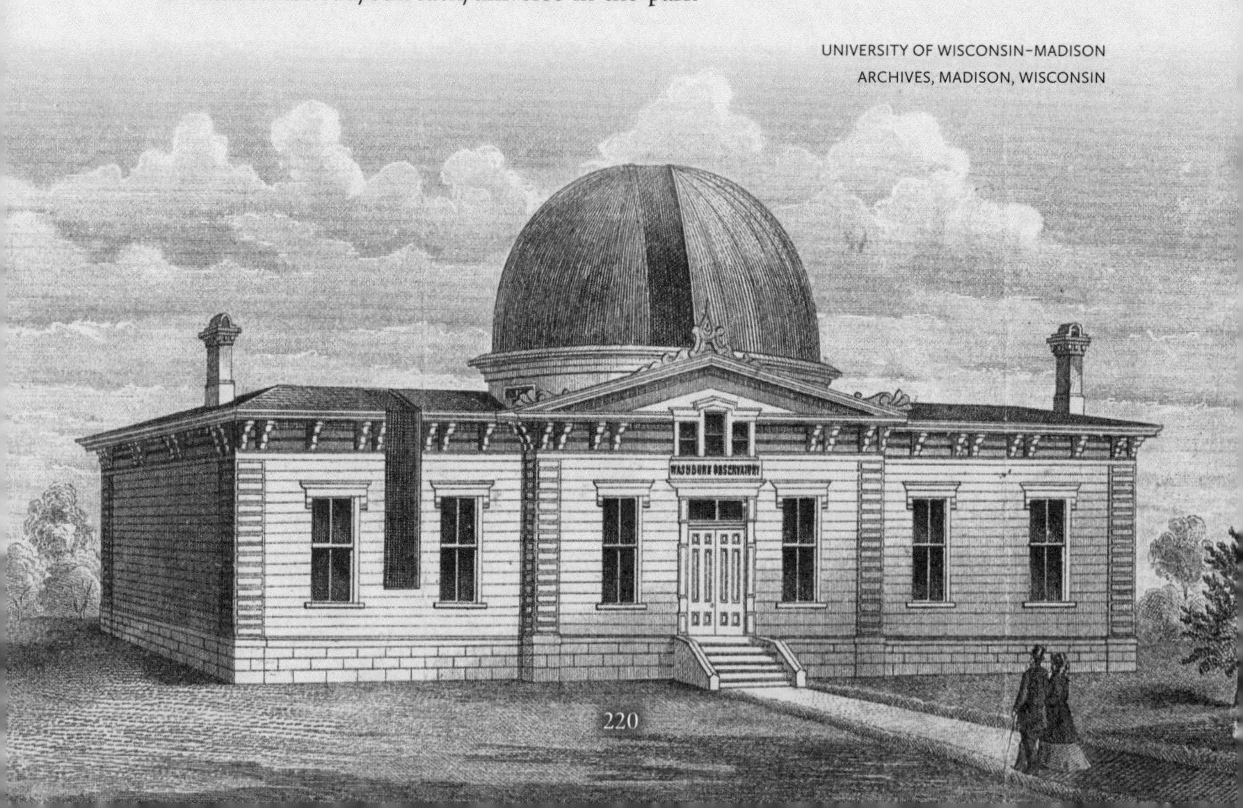

NOTE ON SOURCES

The contents of this book are based on research in both archival and published sources, the most important of which are listed in the bibliography. The University of Wisconsin Archives, Department of Special Collections, and Digital Collections, all parts of the University of Wisconsin–Madison Library System, are essential repositories for astronomical history, including manuscript sources such as the correspondence of observatory directors, scientific records, rare publications, photographs, and artifacts, as well as essential administrative documents such as reports of the University of Wisconsin Board of Regents. The holdings of the Wisconsin Historical Society Library and Archives, which preserve important manuscript collections and photographs in abundance, are essential resources as well. Less obvious at first glance, but also very important, owing to the entangled histories of the two observatories, are the Lick Observatory Records, located in Special Collections and Archives of the University of California, Santa Cruz.

The bibliography lists the most important primary and secondary sources for the history of Washburn Observatory, the Space Astronomy Laboratory, the research conducted by astronomers there, and the lives of the people associated in various ways with astronomy at the University of Wisconsin. The notes section contains citations and sources for the direct quotations that appear throughout the book.

Naturally, the *Publications of the Washburn Observatory*, in fifteen volumes, subdivided into many parts, published by the state of Wisconsin between 1882 and 1934, are essential sources, and they can be readily found in physical and digital forms. The volumes, edited by the observatory directors from Holden through Stebbins, contain descriptions of buildings, instruments, and research programs, as well as observational data and scientific results.

The *Biographical Memoirs* of the National Academy of Sciences constitute another important primary source, given that every director of Washburn Observatory was elected to the Academy and the *Memoirs* are commonly written by close associates, often students, of their subjects. Thus, we have a remarkable chain in these *Memoirs* that is very important for Washburn Observatory history: Comstock wrote Watson's *Memoir*, Stebbins wrote Comstock's, Whitford wrote Stebbins's, and Wisconsin astronomer Donald Osterbrock wrote Whitford's. Holden collaborated with Comstock on the Watson *Memoir*, and Holden's was written by W. W. Campbell, his successor as director of Lick Observatory. Code's *Memoir* has not been published at the time of this writing.

For more recent history, the Oral History Interviews of the Niels Bohr Library and Archives of the American Institute of Physics provide essential insights into the development of modern astronomy and astrophysics. Huffer, Whitford, and Code are among the major figures who gave interviews for that project.

The chapter on the earliest stargazers was informed by verbal and written insights from scholars and members of the Dakota and Ojibwe Nations, conveyed in various media, as well as a variety of other scholarly sources.

Published historical sources provide insight and context in many forms. Jim Feldman's *Buildings of the University of Wisconsin* deserves special mention here, as does the *Biographical Encyclopedia of Astronomers* (Thomas Hockey, editor-in-chief), which contains articles on many historically important astronomers.

The photographs reproduced here are taken mostly from the collection of prints, negatives, and slides of the University of Wisconsin–Madison Department of Astronomy. A majority of the rest come from the photographic archives of the Wisconsin Historical Society and the University of Wisconsin Archives. Also included are some captured over the years by photographers in University Communications, including Jeff Miller. A few are used with permission from other archives and collections, as noted.

SELECTED BIBLIOGRAPHY

The biographical memoirs of the National Academy of Science are available online at www.nasonline.org/publications/biographical-memoirs/.

The *Publications of the Washburn Observatory* from 1882 to 1934 are available online at HathiTrust, www.hathitrust.org, and other sources, as well as physically in various libraries.

Sidney Dean Townley's *Diary of a Student of the University of Wisconsin, 1886 to 1892*, is accessible online from the University of Wisconsin–Madison Digital Collections at https://search.library.wisc.edu/digital/ANXOGO2Q4EXQLX8G.

ARCHIVAL SOURCES

Niels Bohr Library and Archives, American Institute of Physics. Oral History Interviews. www.aip.org/history-programs/niels-bohr-library/oral-histories.

University of California, Santa Cruz, Special Collections and Archives. Lick Observatory Records. https://guides.library.ucsc.edu/speccoll/lick-observatory.

University of Wisconsin Archives. Astronomy Department Records, 1867–1986. www.library.wisc.edu/archives/.

Wisconsin Historical Society. Archives and photograph collection. www.wisconsinhistory.org.

PUBLISHED SOURCES

Baron, David. *American Eclipse: A Nation's Epic Race to Catch the Shadow of the Moon and Win the Glory of the World*. New York: Liveright, 2018.

Bartky, Ian R. *Selling the True Time: Nineteenth-Century Timekeeping in America*. Stanford, CA: Stanford University Press, 2000.

Baum, Richard, and William Sheehan. *In Search of Planet Vulcan: The Ghost in Newton's Clockwork Universe*. New York: Plenum Press, 1997.

Birmingham, Robert A. *Spirits of the Earth: The Effigy Mound Landscape of Madison and the Four Lakes*. Madison: University of Wisconsin Press, 2010.

Buenker, John D. *The History of Wisconsin, Volume IV: The Progressive Era, 1893–1914*. Madison: State Historical Society of Wisconsin, 1998.

Campbell, W. W. "Biographical Memoir of Edward Singleton Holden, 1846–1914." National Academy of Sciences, vol. 8 (1916): 345–372.

City of St. Paul Minnesota, Indian Mounds Regional Park Cultural Landscape Study. *Indian Mounds, A Sacred Place of Burial: Cultural Landscape and Messaging Plan*. June 2020. www.stpaul.gov/sites/default/files/2020-12/IndianMounds_CLS_MP_Final.pdf.

Code, Arthur D. "Photocells; Hot Stars and Spiral Arms." In *Massive Stars: Their Lives in the Interstellar Medium*, Astronomical Society of the Pacific Conference Series, vol. 35 (1993): 3–12. Edited by Joseph P. Cassinelli and Edward B. Churchwell. https://ui.adsabs.harvard.edu/abs/1993ASPC...35....3C/abstract.

Comstock, George C. "Memoir of James Craig Watson, 1838–1880." National Academy of Sciences (1888): 43–57.

Curti, Merle. *The University of Wisconsin: A History*. Madison: University of Wisconsin Press, 1949.

Feldman, Jim. *The Buildings of the University of Wisconsin*. Madison: The University Archives, 1997.

Foucault, Bob, dir. *Northern Nights, Starry Skies*. St. Paul, MN: Public Broadcasting Service North, 2023. www.pbs.org/show/northern-nights-starry-skies/.

Gainor, Christopher. *Not Yet Imagined: A Study of the Hubble Space Telescope Operations*. Washington, DC: NASA Office of Communications, 2020.

Gara, Larry. *Westernized Yankee: The Story of Cyrus Woodman*. Madison: State Historical Society of Wisconsin, 1956.

Hearnshaw, J. B. *The Measurement of Starlight: Two Centuries of Astronomical Photometry*. Cambridge: Cambridge University Press, 1996.

Hockey, Thomas A., et al., eds. *The Biographical Encyclopedia of Astronomers*. Springer Reference. New York: Springer, 2007.

Hoeveler, J. David. *John Bascom and the Origins of the Wisconsin Idea*. Madison: University of Wisconsin Press, 2016.

Holden, Edward S., et al., eds. *Publications of the Washburn Observatory*. Madison: Various state printers, 1882–1934.

Kelsey, Kerck. *Remarkable Americans: The Washburn Family*. Gardiner, ME: Tilbury House, 2008.

Kron, Gerald E. "Obituary: Joel Stebbins, 1878–1966." *Publications of the Astronomical Society of the Pacific* 78 (June 1, 1966): 214–22.

Lattis, Anthony. "Constructing the Electric Eye: Situating the Smithsonian National Air and Space Museum's Wisconsin Collection of Photoelectric Detectors in Historical Context." *Journal for the History of Astronomy* 51, no. 4 (November 17, 2020): 423–60.

Lattis, James. "The Stebbins Galaxy: The Origins of Interstellar Medium Studies in the Shrinking Super-Galaxy." *Journal of Astronomical History and Heritage* 17 (November 1, 2014): 240–57.

———. "Wanderings of the 'Simply Perfect' Burnham Telescope." *Journal of Astronomical History and Heritage* 20 (August 1, 2017): 177–94.

Lee, Annette S. "Native Skywatchers and the Ojibwe Giizhig Anung Masinaaigan—Ojibwe Sky Star Map." *Astronomical Society of the Pacific Conference Series*, vol. 473 (2013).

Liebl, David S., and Christopher Fluke. "Investigations of the Interstellar Medium at Washburn Observatory 1930–58." *Journal of Astronomical History and Heritage* 7 (December 1, 2004): 85–94.

Marché, Jordan D., II, and Adam J. Walsh. "The Wisconsin Experiment Package (WEP) Aboard the Orbiting Astronomical Observatory (OAO-2)." *Journal of Astronomical History and Heritage* 9 (December 1, 2006): 185–99.

Minnesota Historical Society, "The Dakota People." www.mnhs.org/fortsnelling/learn/native-americans/dakota-people.

Nesbit, Robert C. *The History of Wisconsin: Urbanization and industrialization, 1873–1893*. Madison: State Historical Society of Wisconsin, 1985.

Osterbrock, Donald E. "Albert Edward Whitford, 1905–2002, A Biographical Memoir." National Academy of Sciences, vol. 85 (2004): 1–28.

———. "California-Wisconsin Axis in American Astronomy I." *Sky and Telescope* 51 (January 1, 1976): 9–14.

———. "California-Wisconsin Axis in American Astronomy II." *Sky and Telescope* 51 (February 1, 1976): 91.

———. "Further Links in the California-Wisconsin Astronomical Connection." *Transactions of the Wisconsin Academy of Sciences, Arts, and Letters* 69 (1981): 153–162.

———. "The Rise and Fall of Edward S. Holden—Part One." *Journal for the History of Astronomy* 15 (June 1, 1984): 81–127.

———. "The Rise and Fall of Edward S. Holden—Part Two." *Journal for the History of Astronomy* 15 (October 1, 1984): 151–76.

———. *Yerkes Observatory, 1892–1950: The Birth, Near Death, and Resurrection of a Scientific Research Institution*. Chicago: University of Chicago Press, 1997.

Osterbrock, Donald E., John R. Gustafson, and W. J. Shiloh Unruh. *Eye on the Sky: Lick Observatory's First Century*. Berkeley: University of California Press, 1988.

Pauketat, Timothy R., Susan M. Alt, and Jeffrey D. Krutchen. "The Emerald Acropolis: Elevating the Moon and Water in the Rise of Cahokia." *Antiquity* 91, issue 355 (January 20, 2017). https://doi.org/10.15184/aqy.2016.253.

Pauketat, Timothy R., Robert F. Boszhardt, and Michael Kolb. "Trempealeau's Little Bluff: An Early Cahokian Terraformed Landmark in the Upper Mississippi Valley." *Midcontinental Journal of Archaeology* 42, no. 2 (July 5, 2017): 168–99. https://doi.org/10.1080/01461109.2017.1346213.

See, T. J. J. "Brief Biographical Notice of Professor Milton Updegraff." *Publications of the Astronomical Society of the Pacific* 50 (December 1, 1938): 332–34.

Smith, Bruce D. "Mississippian Elites and Solar Alignments: A Reflection of Managerial Necessity, or Levers of Social Inequality?" *Archaeological Papers of the American Anthropological Association* 3, no. 1 (January 1992). https://doi.org/10.1525/ap3a.1992.3.1.11.

Smith, Robert W. *The Space Telescope: A Study of NASA, Science, Technology, and Politics*. Cambridge: Cambridge University Press, 1993.

Stebbins, Joel. "Albert Stowell Flint." *Popular Astronomy* 31 (1923): 369.

———. "Biographical Memoir of George Cary Comstock, 1855–1934." National Academy of Sciences, vol. 20 (1938): 159–182.

———. "Jakob Kunz." *Popular Astronomy* 47 (March 1, 1939): 117–21.

———. "Washburn Observatory, 1878–1958." *Publications of the Astronomical Society of the Pacific* 70 (1958): 437–449.

Stebbins, Joel, and Albert E. Whitford. "Diameter of the Andromeda Nebula." *Contributions from the Mount Wilson Observatory* 3 (1934): 67–72.

Thwaites, Reuben Gold. *The University of Wisconsin: Its History and Its Alumni: With Historical and Descriptive Sketches of Madison*. Madison: J. N. Purcell, 1900.

Whitesell, Patricia S. *A Creation of His Own: Tappan's Detroit Observatory*. Ann Arbor: Michigan Historical Collections, Bentley Historical Library, University of Michigan, 1998.

———. "Detroit Observatory: Nineteenth-Century Training Ground for Astronomers." *Journal of Astronomical History and Heritage* 6 (December 1, 2003): 69–106.

Whitford, Albert. "Joel Stebbins, 1878–1966: Biographical Memoir." National Academy of Sciences, (1978): 291–316.

Wisconsin Historical Society, "Ancient Land and First Peoples." www.wisconsinhistory.org/Records/Article/CS3584.

NOTES

CHAPTER 1

"When I did that, that's when things started to go together": Carl Gawboy, interview. Foucault, *Northern Nights, Starry Skies*.

"Given the myriad of possible sighting points and lines among the thousands of differently oriented mounds and their appendages": Birmingham, *Spirits of the Earth*, 41.

"These are striking memorials": *Publications of the Washburn Observatory*, vol. 1, 5–6.

CHAPTER 3

"Equal or superior to that of the Observatory of Harvard University, Cambridge": *Wisconsin State Journal*, September 18, 1877.

CHAPTER 7

"an extremely interesting man" whose "conversation was entertaining to the point of brilliancy": Campbell, "Biographical Memoir of Edward Singleton Holden," 357.

CHAPTER 12

"We are having a very snowy winter and it has been very amusing to observe": Alice Lamb to Edward Holden, January 21, 1886, Lick Observatory Archives, Correspondence: Washburn Observatory (Alice M. Lamb) 1885–1889, Box 144, Special Collections and Archives of the University of California, Santa Cruz.

"I refused the offer partly because I had far rather work in a true observatory": Alice Lamb to Edward Holden, December 19, 1885, Lick Observatory Archives, Correspondence: Washburn Observatory (Alice M. Lamb) 1885–1889, Box 144, Special Collections and Archives of the University of California, Santa Cruz.

CHAPTER 14

"In his second series of parallax measurements": Annual Report by Albert Flint to George Comstock, 1908, Washburn Observatory General Correspondence, Box 8, 7/4/2, University of Wisconsin–Madison Archives, Madison, Wisconsin.

CHAPTER 17

"I was interested to see how I could manipulate the Bamberg transit after some twenty years": Joel Stebbins to George Comstock, October 17, 1922, Washburn Observatory General Correspondence, Box 10, 7/4/2, University of Wisconsin–Madison Archives, Madison, Wisconsin.

CHAPTER 19

"I went down the Bright Angel Trail": Jakob Kunz to Joel Stebbins, October 13, 1923, Washburn Observatory General Correspondence, Box 10, 7/4/2, University of Wisconsin–Madison Archives, Madison, Wisconsin.

CHAPTER 27

"We shrunk the universe!": Interview with Dr. Charles Huffer by David DeVorkin, July 8, 1977, Niels Bohr Library and Archives, American Institute of Physics. www.aip.org/history/ohilist/4684.html.

CHAPTER 28

"When the size of our galaxy": Stebbins and Whitford, "Diameter of the Andromeda Nebula," 72.

CHAPTER 29

"Once when I was checking my car across the state line into California": Stebbins, "Washburn Observatory," 443.

"I don't know how much trouble": Joel Stebbins to A. F. Gallistel, May 17, 1937, Washburn Observatory General Correspondence, Box 10, 7/4/2, University of Wisconsin–Madison Archives, Madison, Wisconsin.

CHAPTER 31

"As long as you are doing research you can expect to be in difficulties": Stebbins, "Washburn Observatory," 449.

CHAPTER 32

"of the nights to come": Stebbins, "Washburn Observatory," 449.

CHAPTER 36

"watching the dancing beam of light on the galvanometer scale": Code, "Photocells; Hot Stars and Spiral Arms," 6.

"eases our consciences here on taking Albert from Wisconsin": Joel Stebbins to Olin Eggen, December 18, 1957, Lick History: Astronomers: Stebbins, Joel—Isabelle Dodge gift, 1934–1956, Lick Observatory Archives, Special Collections and Archives of the University of California, Santa Cruz.

SPOTLIGHT: WISCONSIN'S ASTRONAUTS

"I have seen some incredible sights: lightning spreading over the Pacific": Remarks by NASA Administrator Sean O'Keefe, Laurel Clark Auditorium Dedication Ceremony, July 11, 2003, Bethesda, MD, www.nasa.gov/audience/formedia/speeches/ok_laurel_clark_07.11.03.html.

SPOTLIGHT: WELCOMING THE PUBLIC

"On one of the days during this rather hectic astronomical period": Stebbins, "Biographical Memoir of George Cary Comstock," 174.

"The best environment in which to introduce the general public to astronomy is outside under dark skies": "University in the Park Takes Astronomy to Wisconsin State Parks," University Communications, https://news.wisc.edu/universe-in-the-park-takes-astronomy-to-wisconsin-state-parks.

INDEX

Page numbers in ***bold italics*** refer to images.

ABOUT THE AUTHORS

JEFF MILLER, UNIVERSITY COMMUNICATIONS

James Lattis holds a PhD in History of Science from the University of Wisconsin–Madison. He specializes in early modern astronomy and the history of Wisconsin astronomy and has published books and articles on both. Lattis helped the University of Wisconsin's Space Astronomy Laboratory create UW Space Place and has directed it since its founding. Lattis helps manage the historic Washburn Observatory, teaches courses in introductory astronomy and its history, gives frequent public talks, and consults widely for the media.

ALTHEA DOTZOUR, UNIVERSITY COMMUNICATIONS

Kelly Tyrrell began her career looking through a microscope as a cellular and molecular biologist. She later embarked on a path as a science writer and developed a strong fascination with astronomy. In 2017, Tyrrell traveled to the Southern African Large Telescope for the UW–Madison storytelling project, *Origins*. Today, Tyrrell continues to work as a science writer and oversees media relations for UW–Madison. She also serves as engagement editor for the science journalism nonprofit *The Open Notebook*.